Lecture Notes in Computer Science 14767

Founding Editors

Gerhard Goos
Juris Hartmanis

The series Lecture Notes in Computer Science (LNCS), including its subseries Lecture Notes in Artificial Intelligence (LNAI) and Lecture Notes in Bioinformatics (LNBI), has established itself as a medium for the publication of new developments in computer science and information technology research, teaching, and education.

LNCS enjoys close cooperation with the computer science R & D community, the series counts many renowned academics among its volume editors and paper authors, and collaborates with prestigious societies. Its mission is to serve this international community by providing an invaluable service, mainly focused on the publication of conference and workshop proceedings and postproceedings. LNCS commenced publication in 1973.

Gunel Jahangirova · Foutse Khomh
Editors

Search-Based Software Engineering

16th International Symposium, SSBSE 2024
Porto de Galinhas, Brazil, July 15, 2024
Proceedings

 Springer

Editors
Gunel Jahangirova
King's College London
London, UK

Foutse Khomh
Polytechnique Montréal
Montréal, QC, Canada

ISSN 0302-9743   ISSN 1611-3349 (electronic)
Lecture Notes in Computer Science
ISBN 978-3-031-64572-3   ISBN 978-3-031-64573-0 (eBook)
https://doi.org/10.1007/978-3-031-64573-0

This Springer imprint is published by the registered company Springer Nature Switzerland AG
The registered company address is: Gewerbestrasse 11, 6330 Cham, Switzerland

If disposing of this product, please recycle the paper.

Preface

A Message from the General and Program Chairs

Welcome to the proceedings of the 16th Symposium on Search-Based Software Engineering, SSBSE 2024, held in Porto de Galinhas, Brazil. The symposium was co-located with the ACM International Conference on the Foundations of Software Engineering, which is one of the main conferences in the field of software engineering. SSBSE 2024 featured four tracks: Research, Challenge, Hot Off the Press, and the Replication & Negative Results and New Ideas & Emerging Results (RENE/NIER) track.

Each of the tracks had a The Hot Off the Press (HOP) track offered authors of recently published papers the opportunity to present their work to the SSBSE community by giving a talk at the conference. The Replications and Negative Results Track provided a venue for researchers to submit replications of all types of empirical studies related to Search-Based Software Engineering, and original works reporting negative results on any of the topics of interest for the research track of the SSBSE conference. The Challenge Track was an exciting opportunity for SBSE researchers to apply tools, techniques, and algorithms to real-world software. Participants used their expertise to carry out analyses on open-source software projects or to directly improve the infrastructure powering research experiments. The principal criterion for the challenge track was to produce interesting results and to apply expertise to challenge the state of the art and inspire future SBSE research. The challenges for this year were on quantum computing and generative AI & SSBSE.

The research track received 7 submissions. To ensure an unbiased evaluation process, we adopted double-anonymous reviewing for this track. Each paper was assigned three reviewers and went through a thorough review process. As a result, 3 submissions out of 7 (43%) were accepted for the research track. The challenge track received 7 submissions and 4 of them (57%) were accepted. The HOP track accepted 2 submissions out of 3 (66%) and the RENE/NIER track accepted 1 out of 2 (50%). Overall, SSBSE 2024 accepted 10 papers out of 19 (53%). The decisions on the papers were exclusively derived from the Program Committees' deliberations on the content and quality of each paper, with no regard for quotas.

Many people participated in the organisation of SSBSE 2024 and the preparation of the proceedings. The HOP track was co-chaired by Vesna Nowack (Imperial College London, UK) and Vincenzo Riccio (University of Udine, Italy). The RENE/NIER track was co-chaired by Gabin An (Korea Advanced Institute of Science and Technology, South Korea) and Matheus Paixao (State University of Ceará, Brazil). The challenge track was handled by co-chairs Karine Even-Mendoza (King's College London, UK), Hector Menendez (King's College London, UK), and Harel Berger (Georgetown University, USA). As part of the challenge track, together with University College London Crest Centre, they organised a collaborative jam session that was open to the public and ran from March 18th to 19th. We would like to thank the co-chairs and Program Committee of all tracks, as well as the authors of all submissions, for their invaluable contribution to SSBSE 2024.

Last, but not least, we want to thank all the members of the SSBSE community for attending and participating in SSBSE 2024! We hope you had a great time at Porto de Galinhas and enjoyed the program. If you couldn't attend, our proceedings are here for you.

July 2024

Marcio Barros
Gunel Jahangirova
Foutse Khomh

Organization

General Chair

Márcio Barros   Federal University of the State of Rio de Janeiro (UNIRIO), Brazil

Program Chairs

Gunel Jahangirova   King's College London, UK
Foutse Khomh   Polytechnique Montréal, Canada

Research Track Chairs

Gunel Jahangirova   King's College London, UK
Foutse Khomh   Polytechnique Montréal, Canada

Research Track Program Committee

Shaukat Ali	Simula Research Laboratory, Norway
	Oslo Metropolitan University, Norway
Paolo Arcaini	National Institute of Informatics, Japan
Andrea Arcuri	Kristiania University College, Norway
	Oslo Metropolitan University, Norway
Aitor Arrieta	Mondragon University, Spain
Matteo Biagiola	Università della Svizzera italiana, Italy
Thiago Ferreira	University of Michigan - Flint, USA
Gordon Fraser	University of Passau, Germany
Alessio Gambi	IMC University of Applied Sciences Krems, Austria
Gregory Gay	Chalmers, University of Gothenburg, Sweden
Lars Grunske	Humboldt-Universität zu Berlin, Germany
Hadi Hemmati	York University, Canada
Nargiz Humbatova	USI Lugano, Switzerland
Fuyuki Ishikawa	National Institute of Informatics, Japan
Fitsum Kifetew	Fondazione Bruno Kessler, Italy

Phil McMinn	University of Sheffield, UK
Inmaculada Medina-Bulo	Universidad de Cádiz, Spain
Manuel Núñez	Universidad Complutense de Madrid, Spain
Vincenzo Riccio	University of Udine, Italy
José Miguel Rojas	University of Sheffield, UK
Valerio Terragni	University of Auckland, New Zealand
Silvia Vergilio	Federal University of Paraná, Brazil

Challenge Track Chairs

Karine Even-Mendoza	King's College London, UK
Hector Menendez	King's College London, UK
Harel Berger	Georgetown University, USA

Challenge Track Program Committee

Kate M. Bowers	Oakland University, USA
Alexander Brownlee	University of Stirling, UK
José Campos	University of Porto, Portugal
David Clark	University College London, UK
Sophie Fortz	King's College London, UK
Gregory Kapfhammer	Allegheny College, USA
William Langdon	University College London, UK
Stephan Lukasczyk	University of Passau, Germany
Phil McMinn	University of Sheffield, UK
Inmaculada Medina-Bulo	Universidad de Cádiz, Spain
Mohammad Reza Mousavi	King's College London, UK
Justyna Petke	University College London, UK
José Miguel Rojas	University of Sheffield, UK
Dominik Sobania	Johannes Gutenberg University Mainz, Germany
Vali Tawosi	J.P. Morgan AI Research, UK

Hot Off the Press Track Chairs

| Vesna Nowack | Imperial College London, UK |
| Vincenzo Riccio | University of Udine, Italy |

Hot Off the Press Track Program Committee

Aitor Arrieta	Mondragon University, Spain
Matteo Biagiola	Università della Svizzera italiana, Italy
Alexander Brownlee	University of Stirling, UK
Neelofar Neelofar	Monash University, Australia
Fabrizio Pastore	University of Luxembourg, Luxembourg
Donghwan Shin	University of Sheffield, UK
Silvia Vergilio	Federal University of Paraná, Brazil
Tahereh Zohdinasab	USI Lugano, Switzerland

RENE/NIER Track Chairs

Gabin An	Korea Advanced Institute of Science and Technology, South Korea
Matheus Paixão	State University of Ceará, Brazil

RENE/NIER Track Program Committee

Allysson Allex Araújo	Federal University of Cariri, Brazil
Andrea Arcuri	Kristiania University College, Norway Oslo Metropolitan University, Norway
Thelma E. Colanzi Lopez	State University of Maringá, Brazil
Thiago Ferreira	University of Michigan - Flint, USA
Erik Fredericks	Grand Valley State University, USA
Carol Hanna	University College London, UK
Chaiyong Ragkhitwetsagul	Mahidol University, Thailand
José Raúl Romero	University of Cordoba, Spain
Jeongju Sohn	Kyungpook National University, South Korea
Shin Yoo	Korea Advanced Institute of Science and Technology, South Korea

Publicity Chair

Antony Bartlett	TU Delft, Netherlands

Steering Committee

Shaukat Ali	Simula Research Laboratory, Norway
Andrea Arcuri	Kristiania University College, Norway
	Oslo Metropolitan University, Norway
Federica Sarro	University College London, UK
Giovani Guizzo	Brick Abode, Brazil
Gordon Fraser	University of Passau, Germany
Justyna Petke	University College London, UK
Phil McMinn	University of Sheffield, UK
Shin Yoo	Korea Advanced Institute of Science and
	Technology, South Korea
Silvia Vergilio	Federal University of Paraná, Brazil

Contents

Evolutionary Analysis of Alloy Specifications with an Adaptive Fitness Function

Jianghao Wang[1]([⊠])[iD], Clay Stevens[2][iD], Brooke Kidmose[3][iD],
Myra B. Cohen[2][iD], and Hamid Bagheri[1][iD]

[1] University of Nebraska-Lincoln, Lincoln, NE 68588, USA
{jwang65,bagheri}@unl.edu
[2] Iowa State University, Ames, IA 50011, USA
{cdsteven,mcohen}@iastate.edu
[3] Technical University of Denmark, Kongens Lyngby, Denmark
blam@dtu.dk

Abstract. The use of formal methods in software engineering imparts a high degree of rigor and precision on the software development process. While formal methods are crucial for ensuring system dependability, their practical adoption has been limited in part due to scalability concerns, even though many automated analysis tools are available. In this paper, we address the scalability challenge in one type of formal analysis approach, model-finding. Prior work on EvoAlloy has demonstrated the potential for extending the Alloy Analyzer with an evolutionary algorithm by loosening the completeness guarantee while preserving soundness. However, that approach was evaluated on a small set of programs and failed to find many small-scope models that Alloy can find. In this work we introduce a new technique, called AdaptiveAlloy, which uses a novel adaptive fitness function for the analysis of Alloy relational logic specifications. Through our experiments, we illustrate that AdaptiveAlloy is capable of finding models of higher scope, and achieving greater scalability than both EvoAlloy and a state-of-the-art Alloy analyzer.

Keywords: Genetic Algorithm · Formal Analysis · Adaptive Fitness

1 Introduction

Software engineers rely on a wide variety of tools to help them develop secure, efficient, and dependable software. In many software engineering domains—particularly for systems where reliable execution is paramount—developers employ *formal analysis* to verify their systems behave as expected. Researchers have developed and refined a variety of formal approaches to analyze software, applying their tools to validate software in domains such as autonomous vehicles [17], the Internet-of-Things [2], and database design [19,27]. Unfortunately, despite great advances in these approaches, formal analysis techniques still face challenges regarding scalability when applied to large-scale software systems. Modern analysis tools—e.g., Alloy [12]—attempt to address these challenges by

G. Jahangirova and F. Khomh (Eds.): SSBSE 2024, LNCS 14767, pp. 1–17, 2024.
https://doi.org/10.1007/978-3-031-64573-0_1

defining *bounds* on the scope of the analysis; these bounded analysis approaches improve scalability for small applications, but larger scopes remain intractable even for state-of-the-art techniques.

These scalability challenges stem in part from a reliance on Boolean satisfiability (SAT) solvers, which many state-of-the-art formal techniques share. These approaches translate a formal specification into a satisfiability problem— typically represented in conjunctive normal form (CNF)—and provide that problem to an off-the-shelf SAT solver to solve. This translation from specification to SAT is a resource intensive process, requiring a great deal of system memory for large specifications. Once translated, the SAT solver systematically explores a vast search space which grows exponentially in the number of variables in the original specification—a total exploration of which may well be intractable.

In cases where the space of possible solutions is simply too large for systematic exploration, *search-based* techniques show promise as a sound alternative. In particular, EvoAlloy [23] proposed the use of a genetic algorithm to address the task of finding satisfying models within the Alloy context, yielding promising results. It represents potential assignments of relational variables in the Alloy specification as the genotype and utilizes methods such as mutation, crossover, and selection to navigate the solution space and ultimately reach an assignment that satisfies all the constraints. While EvoAlloy [23] represents a notable step forward, it primarily focuses on employing the GA to explore the solution space and converge on satisfactory assignments. However, its approach to fitness evaluation, which relies on "maxsat" and considers only the "top-level" subformulas of the specification, might not fully grasp the intricacies of Alloy's relational specifications. This approach could lead to early convergence on locally optimal solutions, potentially missing satisfying models of the specification.

In this paper, we present a novel approach that significantly enhances the capabilities of genetic algorithm-based analysis in the context of Alloy specifications. The key contribution lies in the refinement and depth of the fitness function employed. Unlike EvoAlloy's [23] approach, which operates on "top-level" subformulas, our proposed fitness function delves into the abstract syntax tree (AST) of the relational formula, offering a granular examination of the specification's structure. By traversing the AST and computing the number of genes that would require modification to satisfy the specification, our approach provides a more comprehensive and nuanced evaluation of candidate solutions. This nuanced evaluation enables our method, implemented in our custom tool ADAPTIVEALLOY, to effectively navigate the solution space and converge more quickly on satisfying solutions. Furthermore, we utilize adaptive fitness (e.g. [3]) in this domain, which dynamically adjusts the weighting of subformulas based on their complexity and difficulty in satisfying the specification. This adaptive approach allows ADAPTIVEALLOY to allocate resources more efficiently, focusing computational effort where it is most needed and enhancing the overall effectiveness of the GA-based analysis of Alloy specifications.

Our comparative analysis of ADAPTIVEALLOY with the Alloy Analyzer and EvoAlloy demonstrates scalability and efficiency improvements across various experimental subjects. Our approach shows enhancements in analysis time of up to 182 times faster compared to the Alloy Analyzer and up to 172 times faster compared to EvoAlloy. Empowered by adaptive fitness functions and granular AST assessment, our approach demonstrates promising capabilities in addressing the scalability and performance challenges faced by current state-of-the-art tools.

2 Background and Running Example

In this section, we present a small, yet representative Alloy specification to illustrate our evolutionary search technique and thus motivate our research. An in-depth discussion of our adaptive approach is outlined in Sect. 3.

Alloy specifications consist of a set of *relations*, defined in a syntax akin to object-oriented programming languages, and a set of *constraints*, expressed as first-order logic sentences. These constraints may include transitive closure over the defined relations. Additionally, specifications may contain one or more *commands*, which aim to find models satisfying the constraints or counterexamples, all within a specified *scope* defined on one or more of the relations.

Listing ?? depicts an Alloy specification of a Binary Expression Tree (BET). This specification outlines the BET's primary data types using four distinct signatures (lines 1–6) and enforces several key constraints (lines 8–19). Specifically, it ensures that each node possesses at most one parent, every expression possesses exactly two children, and no expression self-includes. Moreover, the Root expression is designated as exclusive and holds access to all other nodes.

The Alloy Analyzer translates specifications into a finite relational model using Kodkod [22]. This process involves defining bounds for each relation, which encompass possible tuples. Kodkod then converts these relations, bounds, and constraints into a Boolean formula, which is solved by SAT solvers to identify valid instances. However, Alloy's scalability is limited by its reliance on SAT solvers, which employ an exhaustive enumeration approach, hindering its application to real-world systems. In contrast, our approach, ADAPTIVEALLOY, replaces SAT solvers with a genetic algorithm, offering improved scalability and performance. The next section discusses how our approach successfully achieves a more economical and scalable model-finding technique by using a novel adaptive genetic algorithm in detail.

3 AdaptiveAlloy

Figure 1 presents an overview of ADAPTIVEALLOY and elucidates its ability to circumvent the computationally intensive aspects of the current Alloy Analyzer. On the top, the Alloy Analyzer first reads an Alloy specification and converts it into a relational model. This model is then forwarded to Kodkod. Using the scopes and signature bounds provided by Alloy, Kodkod concretizes these parameters to define the specification boundaries. To represent this finite relational model as a Boolean logic formula, Kodkod maps each relation to a Boolean matrix. Within this matrix, every tuple within the bounds of the given relation corresponds to a unique Boolean variable. The relational constraints are then transformed into Boolean constraints over these translated variables. Subsequently, Kodkod translates the resulting Boolean formula into Conjunctive Normal Form (CNF), which is then passed to an off-the-shelf SAT solver to derive a solution. Finally, the Alloy interpreter interprets the SAT solver's output, translating it into a solution instance.

Both EvoAlloy and ADAPTIVEALLOY modify the process of finding satisfying models of a given specification by circumventing the traditional SAT solver-based approach, as depicted in Fig. 1. ADAPTIVEALLOY entails the following steps: (1) It begins by converting the Alloy specification into a bounded relational model, akin to the process employed by Kodkod in traditional analysis methods. (2) Next, it constructs a genotype representation of candidate solutions. This representation encapsulates the assignments of tuples to the relations within the model. (3) The crux of our approach lies in executing a genetic algorithm-based search of the solution space. This search employs crossover, mutation, and selection techniques, backed by our adaptive Alloy-specific fitness function, to iteratively explore and refine potential solutions.

The following subsections detail our approach, focusing on key components: *Genotypic Representation (Sect. 3.1)* elucidates the methodology behind encoding Alloy specifications, required for enabling our genetic algorithm to effectively navigate the solution space. *Genetic Algorithm Processes (Sect. 3.2)* delve into the processes of crossover, mutation, and selection employed by our genetic algorithm, instrumental in iteratively refining and improving candidate solutions.

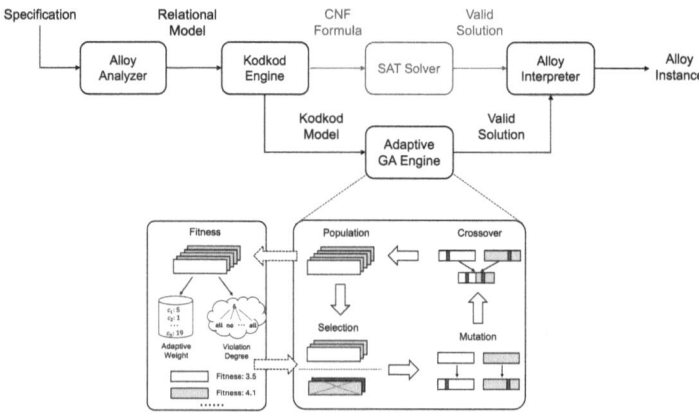

Fig. 1. ADAPTIVEALLOY overview

Following this, our fitness function, which incorporates two core innovations, is presented: *Degree of Violation Computation (Sect. 3.3)* introduces a method for computing the "degree" to which a given assignment violates the Alloy specification. This granular analysis provides valuable insights into the quality of candidate solutions. Lastly, *Dynamic Weight Adjustments (Sect. 3.41)* explains how our dynamic weight adjustments enhance our approach's efficacy in navigating complex solution spaces by allocating more weight to challenging subformulas.

3.1   Genotypic Representation

The bounded relational models used by Kodkod are typically converted into Boolean variables in the underlying SAT problem by creating a unique variable to represent each possible tuple assignment to each relation based on the bounds defined for that relation. If a given variable is true in a given candidate solution, the corresponding tuple *is* assigned to the corresponding relation in that candidate;

$$\varnothing \subseteq \mathsf{Var} \subseteq \{\, V1,\ V2,\ V3\,\}$$

$\mathsf{Var}_{s0} = \{\ \}$	$\Longrightarrow 0\ 0\ 0$
$\mathsf{Var}_{s1} = \{\ V2\ \}$	$\Longrightarrow 0\ 1\ 0$
$\mathsf{Var}_{s2} = \{\ V1,\ V3\ \}$	$\Longrightarrow 1\ 0\ 1$

...

Fig. 2. Example (bit-string) chromosome representation of tuple assignments to relation Var.

if the variable is false, that tuple is *not* assigned to the relation. In ADAPTIVEAL-LOY, we use a similar mapping to represent the genotype for each individual as a set of chromosomes corresponding to the set of relations, where each gene in each chromosome is a single bit value (1 or 0) representing the assignment/non-assignment of a specific tuple to that relation, respectively. Thus, each individual can be defined genetically as a bitstring of genes indicating the assignment/non-assignment of each relation-tuple pair that falls within the bounds of the specification. Figure 2 depicts examples of bit chromosomes created for the Var relation in the example from Sect. 2.

3.2   Genetic Algorithm

For ADAPTIVEALLOY's initial generation, we employ a combination of random gene assignment for the majority of individuals and a domain-specific strategy. This strategy generates two special chromosomes: one composed of all 1 s in a bit-string format, representing an instance incorporating all tuples for each relation, and another composed of all 0 s, representing an empty tuple set instance, thus providing a diversity of alleles in the population. Figure 3 provides an illustration of various aspects of ADAPTIVEALLOY, including its (a) chromosome representation, (b) two arbitrarily selected chromosomes corresponding to Listing ??, (c) transformation from tuple-sets form into bit-string chromosome, (d) crossover step for generating a new bit string, and (e) mutation process over the bit-string.

Selection: ADAPTIVEALLOY's selector employs a combination of *elitism* and *tournament selection* strategies to determine the population for the subsequent generation. Initially, the selector retains the e most-fit chromosomes from the current population, adding them unchanged to the next generation (*elitism*). Subsequently, it randomly selects t individuals from the remaining population and iteratively picks the most-fit individual among those t, repeating this process until all individuals are chosen (*tournament selection*). The next generation consists of a set of *survivors*, comprising the elites (e) and a portion of the winners from the tournament selection, along with a set of *offspring* generated through crossover and mutation. The number of survivors and offspring is determined based on the ratio $rate_s$, with the total population size (p) being composed of these individuals for the subsequent generation.

Crossover: After the mating pool is established, ADAPTIVEALLOY employs the *crossover* operation to produce offspring for the subsequent generation. The

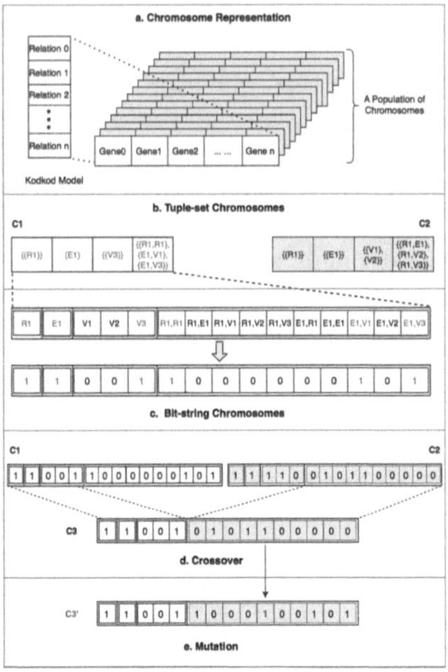

Fig. 3. ADAPTIVEALLOY's (a) chromosome representation, (b) two arbitrarily picked chromosomes for Listing ??, (c) transformation from tuplesets form into bit-string chromosome,(d)crossover step for creating a new bit string, and (d) mutation over bit-string.

crossover process starts by randomly choosing two parent genotypes from the mating pool. Subsequently, a random bit index ranging from 0 to the length of the shorter of the two bit-string genotypes is selected as the *cut point* for *one-point* crossover. Using this cut point, the crossover operator generates two new individuals by exchanging the bits to the right of the cut point between each of the parents.

Mutation: To maintain genetic diversity and prevent premature convergence to local optima, ADAPTIVEALLOY employs mutation as a crucial genetic operator. The mutation process in ADAPTIVEALLOY involves a strategic combination of configurable mutation rates and a probability-based selection of mutation operators. The mutation process operates at two levels: chromosomes and genes within those chromosomes. The mutation rate for chromosomes, denoted as $p_{individual}$, determines the likelihood that a chromosome will be selected for mutation, while the gene mutation rate, denoted as p_{gene}, determines the likelihood that a gene within the selected chromosome will be altered. More specifically, it can be represented mathematically as follows:

$$P(\text{mutation}) = p_{\text{individual}} \times p_{\text{gene}}$$

where $P(\text{mutation})$ represents the probability of mutation occurring.

Moreover, the mutation operator is determined based on a probability distribution. The choice of mutation operator is made from a set of options each with its associated probability. These operators include:

- **Chromosome Creation:** If the selected chromosome has only 0 s assigned to each individual, the creation operator generates a new equal-length bit chromosome by randomly assigning 0 or 1 for each gene in the chromosome.
- **Chromosome Removal:** Each gene in the selected chromosome is replaced with a 0, effectively altering the chromosome's composition.
- **Chromosome Transformation:** The original value of the selected chromosome is replaced with a newly generated bit chromosome.
- **Bit Transformation:** This operator focuses on altering individual genes within the selected chromosome. It randomly selects a gene and flips its value, thereby introducing localized changes.

3.3   Granular Fitness Analysis: Assessing Degree of Violation

Unlike EvoAlloy [23], which relies on high-level assessments of solution quality, our fitness function delves into the details of the relational formula structure, offering a nuanced assessment of candidate chromosomes. Specifically, our fitness function goes beyond simply counting violated constraints, aiming to capture the diversity and complexity of constraint violations.

To facilitate our fitness function's analysis, we categorize relational formulas into two main classes: (1) Elementary Formulas: These include multiplicity, comparison, and int comparison formulas. They represent basic building blocks of the relational formula and can be evaluated directly. (2) Composite Formulas: These connect multiple elementary formulas with logical operators and often depend on the truth values of more than one subformula. Examples include n-ary, binary, and quantified formulas.

We conceptualize the relational formula as a large abstract syntax tree (AST), with the global root symbolizing the conjunction of all subformulas. Figure 4 demonstrates the AST of the relational constraints of the running example. Each leaf node corresponds to an elementary formula, while composite formulas serve as intermediate nodes, connecting smaller subtrees and leaf nodes. Our fitness function conducts a detailed examination of the AST of the relational formula to identify unsatisfied subformulas

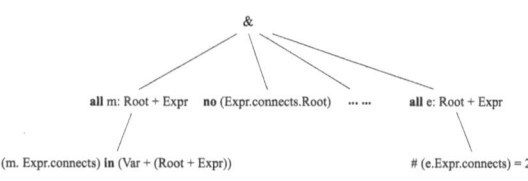

Fig. 4. The Abstract Syntax Tree of the Constraint Formulas

when a chromosome fails to meet the system constraints. This process provides a better understanding of the violated constraints, including the specific subformulas involved and the extent of their violation. For instance, consider a constraint ensuring each node in a graph has exactly two outgoing edges. If a chromosome violates this constraint, we pinpoint the specific subformulas responsible, such as the multiplicity formula expressing the number of outgoing edges for each node.

Chromosomes with different genetic makeup produce varying degrees of violation for the same constraints. Essentially, when a common constraint is evaluated as unsatisfied by distinct chromosomes, the dissimilarity in their genetic makeup often results in the violated tuples varying in both quantity and composition. By analyzing the specific tuples involved in constraint violations, we quantify the degree of violation for each chromosome. An example of this is illustrated in Fig. 5, where chromosomes $C1$ and $C2$ both violate the same constraint, requiring different degrees of modification to satisfy it. This constraint stipulates that for all types of Expr, each one should connect to exactly two child Nodes. Upon evaluating their relational values assigned to Root, Expr, and connects, $C1$ requires only one additional tuple for connects to satisfy this constraint. In contrast, $C2$ has one extra connected node for R1 and is missing two for E1, resulting in a total of three tuples needing revision to meet the constraint.

Based on the detailed analysis of constraint violations and the tuple-wise changes needed for satisfaction, our fitness function computes a fitness score for each chromosome. This score represents the accumulated number of tuple-wise changes required to satisfy all constraints, offering a precise measure of each chromosome's proximity to a valid solution, as represented by the following formula:

$$\sum_{c_i \in Consts} F_t(c_i, ch)$$

where $F_t(c_i, ch)$ indicates the number of violating tuples when evaluating chromosome ch against the ith constraint. This approach precisely quantifies the distance of a specific formula from being satisfied by a chromosome in terms of the number of tuples needed to be altered, in contrast to abstractly counting how many relations are involved in the violation, as considered by prior work. It also supports our design choice of representing the problem using a bit-string chromosome at the tuple level.

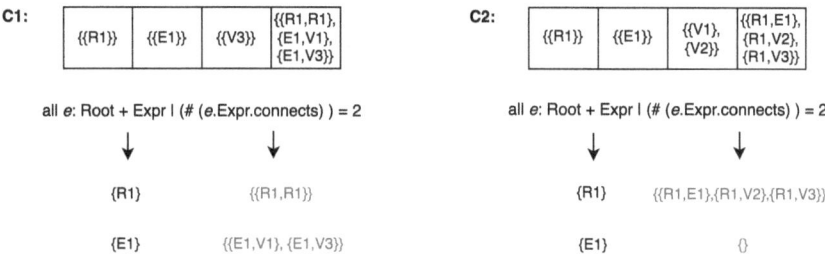

Fig. 5. Two chromosomes with distinct genetic makeup exhibit varying degrees of violation for the same constraint

3.4   Dynamic Weight in Fitness Computation

In addition to tracking the number of tuple revisions required to satisfy each subformula, our approach incorporates a dynamic weight in fitness value computation to tackle the challenge posed by exceptionally difficult constraints. These constraints can often lead to the search converging to local optima, hindering the effectiveness of the genetic algorithm. The adaptive fitness function is defined as follows:

$$w_i = w_i + \Delta w, \quad \Delta w = F_c(c_i, ch*)$$

$$f(ch) = \sum_{c_i \in Consts} w_i \times F_t(c_i, ch)$$

Here, w_i represents the dynamic weight of the ith constraint, which accumulates the number of failed tuples for chromosome ch when the ith constraint is unsatisfied. Initially, all weights are set to 1 and are subsequently updated every certain number of generations by adding given values Δw. $ch*$ denotes the best chromosome in the population found during the period between two consecutive updates of the weights, denoted as P.

The value of Δw is 1 if the best chromosome $ch*$ does not satisfy constraint c_i, and 0 when it is satisfied. This implies that the weights of unsatisfied constraints are increased by 1 periodically. As the population evolves, constraints that persistently remain unsatisfied over extended periods are penalized with higher weights.

When the genetic algorithm encounters a plateau caused by resistant constraints, the adaptive fitness function assigns much lower fitness values to chromosomes that satisfy these constraints. This favors the genes of chromosomes that do not satisfy these constraints to propagate to the next generation, helping to navigate the search out of local optima more efficiently.

This fitness function ensures truth-invariance by requiring the satisfaction of the Alloy specification, which necessitates satisfaction of all its relations and formulas. Our ablation study demonstrates that adaptive fitness outperforms plain fitness significantly, as discussed in detail in Sect. 4.

4   Experimental Evaluation

This section presents the experimental evaluation of ADAPTIVEALLOY. We have implemented ADAPTIVEALLOY's genetic algorithms engine on top of the Alloy Analyzer, its underlying finite relational model finder, Kodkod [21], and the Jenetics framework [26]. ADAPTIVEALLOY consists of two main components: the *Adaptive Evaluator* and the *GA Generator*. The *Adaptive Evaluator* assesses chromosome satisfiability, measures error degrees, and computes adaptive weights for the fitness function. The *GA Generator* produces initial populations, implements mutation operators for effective solution exploration, and facilitates chromosome conversion between Kodkod and bit-string representations. Additionally, it includes a component for transforming chromosome-level model instances into high-level Alloy models at the final stage of the evolutionary search. We used the ADAPTIVEALLOY apparatus for carrying out the experiments. The ADAPTIVEALLOY prototype and data is available on the project website [24]. Our evaluation addresses the following research questions:

- **RQ1.** How does AdaptiveAlloy compare to Alloy and EvoAlloy in terms of both effectiveness and efficiency?
- **RQ2.** What is the impact of Adaptive AST-Based Fitness compared to Non-Adaptive Fitness in terms of performance improvement?

Experimental Subjects. Our experimental subjects consist of publicly available Alloy specifications with varying sizes and complexities. More specifically, we use a list of twelve Alloy specifications modeling prominent algorithms (i.e., Chord models chord protocol for a peer-to-peer distributed hash table) or ubiquitous systems (i.e., Railway models a simplified railway system that declares safety policies for trains) that are distributed with the Alloy Analyzer [1]. When performing the comparison experiments on this collection of specifications, we gradually increased the scope of analysis for each specification. Figure 6 displays the counts of variables and clauses in propositional formulas for each subject system. The data reflects a notable escalation in both variables and clauses as the analysis scope progresses from 5 to 25, highlighting the considerable rise in complexity and computational demands for broader analyses.

Experimental Setup We conducted all experiments on a PC equipped with a 64-core 4.3 GHz AMD Ryzen Threadripper 3990X processor boasting 128 threads and 64 GB of RAM. To maintain consistency, each experiment was allocated 8 cores (16 threads) and 16 GB RAM. Consequently, a maximum of three jobs could run concurrently on the system. Following parameter-tuning, we heuristically settled on the following parameters for all experiments: a population size of 32,

Spec	Analysis Scope			
	5		25	
	Vars	Clauses	Vars	Clauses
abstract-Memory	2,622	4,604	639,497	1,240,899
birthday	2,503	4,245	179,148	335,100
ceilings	952	1,568	44,002	83,578
chord	13,582	32,551	11,643,597	43,930,456
com	9,031	16,447	2,046,548	4,093,018
dijkstra	578	512	63,878	62,552
fileSystem	2,037	3,965	149,946	409,604
grandpa	1,810	2,985	133,594	255,397
handshake	757	1,335	45,214	99,532
life	2,893	8,307	335,508	1,159,923
lists	2,948	8,613	205,940	756,769
railway	2,941	6,078	242,976	922,713

Fig. 6. Size Comparison of Variables and Clauses in Propositional Formulas for Subject Systems Across Two Analysis Scopes of 5 and 25.

with primary GA configurations initialized as follows: 60% offspring fraction, an overall gene mutation rate of 80%, and a one-point crossover with a 60% probability. Additionally, the likelihoods for each mutation operator were configured as follows: for a selected gene represented by a string of 0 s, a 50% chance for bit-string creation and 50% for single-bit creation; for a non-empty selected gene, a 20% chance of deletion, 30% for bit-string transformation, and 50% for single-bit transformation. Regarding the hyper-parameters of our adaptive algorithm, the incremental adaptive weight was initially set to 1, and the adaptive step was set to 100 iterations.

4.1   Results for RQ1: Comparison Against State-of-the-Art

We conducted a comparative analysis of ADAPTIVEALLOY with two state-of-the-art tools: Alloy Analyzer (version 5.1) [1] and EvoAlloy [23]. This comparison aimed to assess how well ADAPTIVEALLOY scales and performs in terms of both effectiveness and efficiency across a range of experimental subjects. We evaluated the effectiveness of ADAPTIVEALLOY by comparing its scalability with Alloy Analyzer and EvoAlloy over increasing analysis scopes. Each technique was subjected to three stopping criteria: reaching a satisfying solution, exceeding the maximum memory allocation, or surpassing a 24-h time limit. To reduce variance, we performed each analysis five times and recorded the analysis time.

Box plots in Fig. 7 illustrate the analysis time (in logarithmic scale) in milliseconds (ms) obtained from Alloy Analyzer, EvoAlloy, AdaptiveAlloy, and AdaptiveAlloy without dynamic weight across increasing analysis scopes for various study objects. The notations "M" and "T" in the diagram denote that the corresponding technique cannot identify a valid solution given the available memory and time resources, respectively; "M" signifies the technique exceeded the maximum memory allocation and "T" indicates the technique surpassed the 24-h time limit. From the experimental results, several observations emerged.

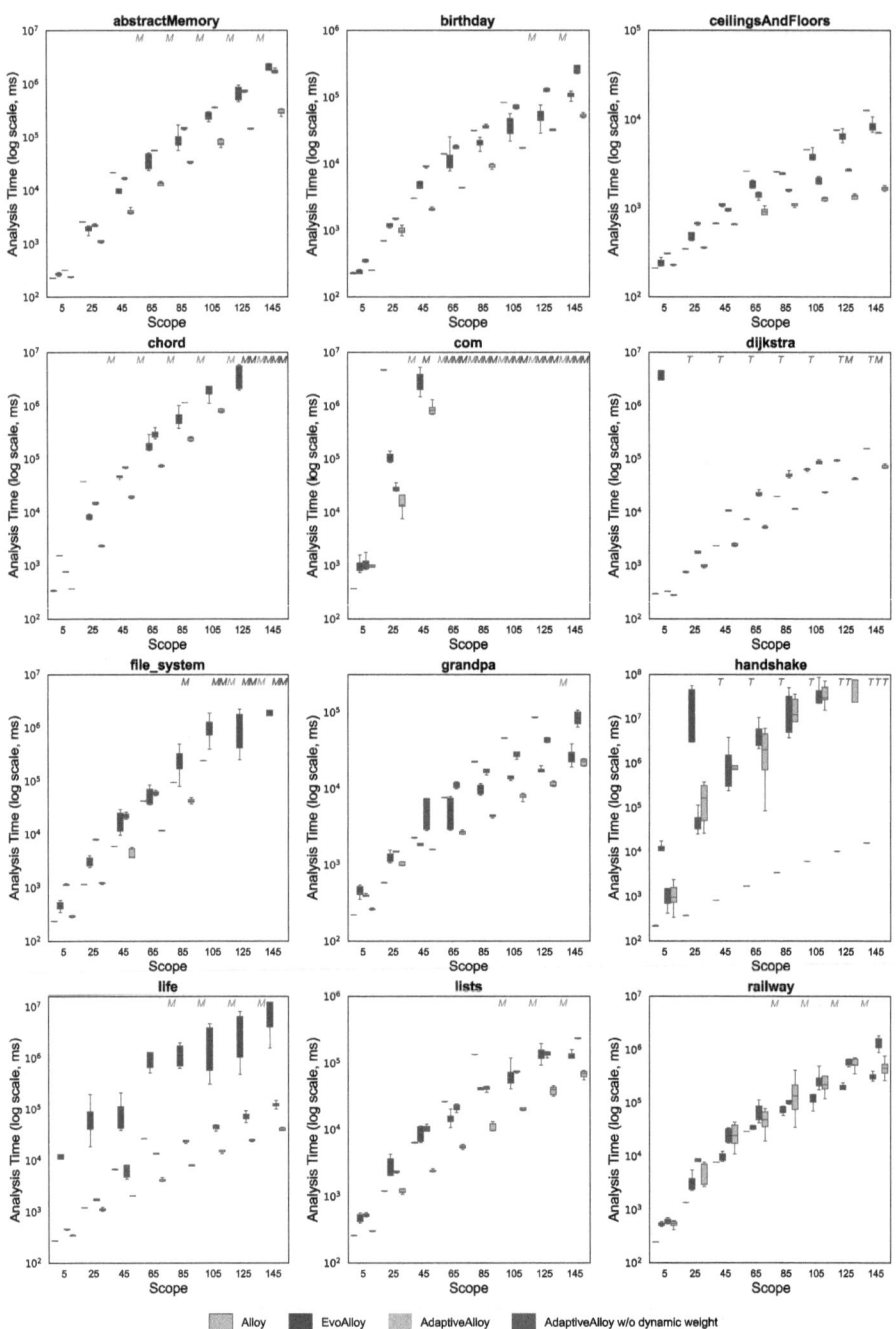

Fig. 7. Box plots depict the analysis time (in logarithmic scale) in milliseconds (ms) taken from Alloy Analyzer, EvoAlloy, AdaptiveAlloy, and AdaptiveAlloy without dynamic weight over the increasing analysis scope across objects of study. **M** denotes exceeding the maximum memory allocation, and **T** indicates surpassing a 24-h time limit.

First, for more than half of the specifications, both EvoAlloy and ADAPTIVEAL-
LOY could scale to larger analysis scopes compared to Alloy Analyzer, which frequently encountered memory limitations. This trend was particularly evident in
specifications such as abstractMemory, birthday, and railway. Second, for smaller
scopes where all three techniques performed well, EvoAlloy generally exhibited
running times comparable to, or worse than, Alloy Analyzer. However, ADAP-
TIVEALLOY outperformed both the Alloy Analyzer and EvoAlloy in terms of the
analysis time required to find solutions. Finally, while EvoAlloy's genetic algorithm (GA) struggled to efficiently solve certain problems, often hitting the time
limit even for small scopes, ADAPTIVEALLOY's advanced adaptive GA approach
proved effective. For instance, in the case of Dijkstra, ADAPTIVEALLOY achieved
superior performance compared to both Alloy Analyzer and EvoAlloy.

> ADAPTIVEALLOY achieves analysis time improvements of up to 181.62
> times faster (with an average enhancement of 20.56 times) compared
> to the Alloy Analyzer and up to 172.10 times faster (with an average
> improvement of 33.04 times) compared to EvoAlloy across various spec-
> ifications.

4.2   Results for RQ2: Ablation Study on Dynamic Weight

To investigate the impact of Adaptive AST-Based Fitness compared to Non-
Adaptive Fitness, we conducted an ablation study on dynamic weight. Our GA-
boosted approach was developed in two phases: initially incorporating advanced
AST-based granular assessment of constraint violation degrees for fitness evaluation, and subsequently adding dynamic weights as indicators of the difficulty
level in satisfying specific subformulas alongside fitness based on constraint violation degrees. Figure 7 outlines the runtime performance of both versions as
the analysis scope increases across the objects of study. The complete form of
ADAPTIVEALLOY(involving both advanced AST-based granular assessment of
constraint violation degree and dynamic weights in fitness analysis) outperforms
the version without introducing dynamic weight for almost every specification
under analysis, with handshake as the only exception, on which both versions
exhibit similar performance. It is noteworthy that, for most specifications, the
efficiency gained from using adaptive fitness becomes increasingly significant as
the analysis scope increases. Notably, for certain specifications like chord and file
system, the non-adaptive version ran out of memory at smaller scopes compared
to the one using adaptive fitness.

> The Ablation Study compared Adaptive AST-Based Fitness with and
> without dynamic weights. Results show dynamic weights significantly
> outperformed static weights, with improvements ranging up to 5.78 times
> (with an average improvement of 3.62 times).

5 Discussion

The experimental results generally indicate that ADAPTIVEALLOY's GA notably improves the scalability of the state-of-the-art without experiencing the runtime efficiency degradation observed in EvoAlloy. However, a few drawbacks of ADAPTIVEALLOY are worth discussing. During the preliminary hyperparameter tuning experiments, we discovered that no universally optimal configuration achieves the best performance across all experimental objects, leaving us a tradeoff option that keeps significantly better running time on the majority of the specs, while maintaining acceptable performance for the rest. This results in ADAPTIVEALLOY having a larger variance in running time over handshake and railway, and performs equally or slightly less efficiently when compared to EvoAlloy. A dynamic parameter tuning technique can potentially further enhance ADAPTIVEALLOY's performance.

ADAPTIVEALLOY enhances the precision of its search guidance by evaluating constraint violation through AST traversal. While this strategy notably boosts fitness function accuracy and overall performance, it comes with increased memory consumption, potentially limiting scalability. Jenetics' memory management shortcomings exacerbate this issue by retaining allocated memory post-iteration. Our ablation study revealed that the non-adaptive ADAPTIVEALLOY variant faced memory constraints due to prolonged search iterations and residual memory accumulation. Implementing a memory-efficient GA engine and imposing a threshold on AST tracking depth could mitigate this challenge.

6 Related Work

Numerous extensions to Alloy and its automated Analyzer have been developed to enhance its performance and address scalability challenges [2, 4, 6, 7, 9, 13–15]. Notable among these are Titanium [5], which optimizes analysis time by generating a complete solution set for original specifications to inform revised ones, and Platinum [28], which partitions constraints into independent subclauses for more efficient analyses. Similarly, iAlloy [25] and SoRBoT [18] leverage solution reuse techniques to enhance efficiency. Aluminum [15] extends the Alloy Analyzer to generate minimal model instances by iteratively removing tuples from found model instances until a minimal instance is reached. Unlike our approach, Aluminum does not incorporate search-based solutions.

EvoAlloy [23] stands out for its focus on scalability, employing evolutionary algorithms to address Alloy Analyzer's limitations. However, its oversimplified problem representation and fitness design hinder its effectiveness. In contrast, our approach, ADAPTIVEALLOY, introduces a sophisticated GA with a novel fitness function and adaptive weight optimization to overcome these limitations.

PLEDGE [16] employs a hybrid metaheuristic search and SMT approach for improving constraint solving, particularly in system testing. While promising, PLEDGE's relies on UML models and OCL constraints. Additionally, PLEDGE lacks significant scalability improvements over Alloy due to its approach of delegating subformulas to an SMT solver, which can be a scalability bottleneck.

In contrast, ADAPTIVEALLOY focuses on improving scalability and efficiency through a Genetic algorithm approach, bypassing intensive solvers.

There is extensive research on using evolutionary algorithms in software engineering [11]. Allmula and Gay [3] propose the use of adaptive fitness functions, however, their focus is on traditional program code coverage. Godefroid and Khurshid apply a genetic algorithm to analyze concurrent reactive systems for errors [10]. ACO-Solver utilizes Ant Colony Optimization for solving intricate string constraints [20]. Concolic Walk combines linear constraint solving with tabu search for complex arithmetic path conditions [8]. In contrast, our work focuses on bounded analysis of large-scale solution spaces specified in relational logic, requiring original chromosome encodings and fitness functions suitable for Alloy's relational logic.

7 Conclusion

In this paper, we introduce a novel approach that enhances genetic algorithm-based analysis, particularly within Alloy specifications. Our key contribution lies in the depth of the fitness function, which offers a granular examination of the specification's structure by traversing the abstract syntax tree. This nuanced evaluation, implemented in our tool, ADAPTIVEALLOY, enables effective navigation of the solution space, leading to globally optimal solutions. Additionally, we introduced an adaptive fitness, dynamically adjusting subformula weighting based on complexity. This optimizes resource allocation, enhancing GA-based analysis efficiency. Our comparative analysis with state-of-the-art Alloy Analyzer and EvoAlloy underscores significant scalability and efficiency improvements, with AdaptiveAlloy achieving analysis times up to 181.62 times faster than Alloy Analyzer and up to 172.10 times faster than EvoAlloy.

For future work, we plan to optimize the memory overhead introduced by the AST traversal tracking procedure as aforementioned. A potential tradeoff can be restricting the maximum depth of the AST for constraints being exploited. Our preliminary parameter tuning results reveal that no single global optimal configuration achieves the best performance for all experimental objects, thus we would also seek to explore incorporating a learning-based technique to dynamically tune the hyperparameters to enhance the performance when analyzing a diversity of specifications.

Acknowledgment. We thank the anonymous reviewers for their valuable comments. This work was supported in part by National Science Foundation awards CCF-1618132, CCF-1755890, CCF-1909688, CCF-2139845, and CCF-2124116.

References

1. Alloy toolset. https://alloytools.org. Accessed Apr 2024
2. Alhanahnah, M., Stevens, C., Bagheri, H.: Scalable analysis of interaction threats in IoT systems. In: ISSTA, pp. 272–285 (2020)

3. Almulla, H., Gay, G.: Learning how to search: generating effective test cases through adaptive fitness function selection. Empir. Softw. Eng. **27**(2), 38 (2022). https://doi.org/10.1007/s10664-021-10048-8
4. Bagheri, H., Kang, E., Malek, S., Jackson, D.: A formal approach for detection of security flaws in the Android permission system. FAOC **30**, 525–544 (2018)
5. Bagheri, H., Malek, S.: Titanium: efficient analysis of evolving alloy specifications. In: FSE, pp. 27–38 (2016)
6. Bagheri, H., Wang, J., Aerts, J., Malek, S.: Efficient, evolutionary security analysis of interacting Android apps. In: ICSME, pp. 357–368. IEEE (2018)
7. Brunel, J., Chemouil, D., Cunha, A., Macedo, N.: The electrum analyzer: model checking relational first-order temporal specifications. In: ASE (2018)
8. Dinges, P., Agha, G.A.: Solving complex path conditions through heuristic search on induced polytopes. In: Proceedings of FSE, pp. 425–436 (2014)
9. Galeotti, J.P., Rosner, N., López Pombo, C.G., Frias, M.F.: Analysis of invariants for efficient bounded verification. In: ISSTA, pp. 25–36 (2010)
10. Godefroid, P., Khurshid, S.: Exploring very large state spaces using genetic algorithms. Int. J. Softw. Tools Technol. Transf. **6**(2), 117–127 (2004)
11. Harman, M., Mansouri, S.A., Zhang, Y.: Search-based software engineering: trends, techniques and applications. ACM Comput. Surv. **45**(1), 11:1–11:61 (2012)
12. Jackson, D.: Software Abstractions, 2nd edn. MIT Press (2012)
13. Milicevic, A., Near, J.P., Kang, E., Jackson, D.: Alloy*: a general-purpose higher-order relational constraint solver. FMSD **55**, 1–32 (2019)
14. Mirzaei, N., Garcia, J., Bagheri, H., Sadeghi, A., Malek, S.: Reducing combinatorics in GUI testing of Android applications. In: ICSE, pp. 559–570 (2016)
15. Nelson, T., Saghafi, S., Dougherty, D.J., Fisler, K., Krishnamurthi, S.: Aluminum: principled scenario exploration through minimality. In: ICSE, pp. 232–241 (2013)
16. Soltana, G., Sabetzadeh, M., Briand, L.C.: Practical constraint solving for generating system test data. TOSEM **29**(2), 11:1–11:48 (2020). https://doi.org/10.1145/3381032
17. Stevens, C., Bagheri, H.: Reducing run-time adaptation space via analysis of possible utility bounds. In: ICSE, pp. 1522–1534. ACM (2020). https://doi.org/10.1145/3377811.3380365
18. Stevens, C., Bagheri, H.: Combining solution reuse and bound tightening for efficient analysis of evolving systems. In: ISSTA, pp. 89–100 (2022)
19. Stevens, C., Bagheri, H.: Parasol: efficient parallel synthesis of large model spaces. In: ESEC/FSE, pp. 620–632. ACM (2022). https://doi.org/10.1145/3540250.3549157
20. Thomé, J., Shar, L.K., Bianculli, D., Briand, L.C.: Search-driven string constraint solving for vulnerability detection. In: ICSE, pp. 198–208 (2017)
21. Torlak, E.: A constraint solver for software engineering: finding models and cores of large relational specifications. Ph.D. thesis, MIT, February 2009
22. Torlak, E., Jackson, D.: Kodkod: a relational model finder. In: Grumberg, O., Huth, M. (eds.) TACAS 2007. LNCS, vol. 4424, pp. 632–647. Springer, Heidelberg (2007). https://doi.org/10.1007/978-3-540-71209-1_49
23. Wang, J., Bagheri, H., Cohen, M.B.: An evolutionary approach for analyzing alloy specifications. In: ASE, pp. 820–825 (2018)
24. Wang, J., Stevens, C., Kidmose, B., Cohen, M.B., Bagheri, H.: AdaptiveAlloy webpage, May 2024. https://sites.google.com/view/adaptivealloy
25. Wang, W., Wang, K., Gligoric, M., Khurshid, S.: Incremental analysis of evolving alloy models. In: Vojnar, T., Zhang, L. (eds.) TACAS 2019. LNCS, vol. 11427, pp. 174–191. Springer, Cham (2019). https://doi.org/10.1007/978-3-030-17462-0_10

26. Wilhelmstötter, F.: Jenetics (2021). http://jenetics.io
27. Wu, N., Simpson, A.C.: Formal relational database design: an exercise in extending the formal template language. FAOC **26**(6), 1231–1269 (2014). https://doi.org/10.1007/S00165-014-0299-6
28. Zheng, G., Bagheri, H., Rothermel, G., Wang, J.: Platinum: reusing constraint solutions in bounded analysis of relational logic. In: FASE 2020. LNCS, vol. 12076, pp. 29–52. Springer, Cham (2020). https://doi.org/10.1007/978-3-030-45234-6_2

Higher Fault Detection Through Novel Density Estimators in Unit Test Generation

Annibale Panichella$^{(\boxtimes)}$ (ID) and Mitchell Olsthoorn (ID)

Delft University of Technology, Delft, The Netherlands
{a.panichella,m.j.g.olsthoorn}@tudelft.nl

Abstract. Many-objective evolutionary algorithms (MOEAs) have been applied in the software testing literature to automate the generation of test cases. While previous studies confirmed the superiority of MOEAs over other algorithms, one of the open challenges is maintaining a strong selective pressure considering the large number of objectives to optimize (coverage targets). This paper investigates four density estimators as a substitute for the traditional crowding distance. In particular, we consider two estimators previously proposed in the evolutionary computation community, namely the *subvector-dominance assignment* (SD) and the *epsilon-dominance assignment* (ED). We further propose two novel density estimators specific to test case generation, namely the *token-based density estimator* (TDE) and the *path-based density estimator* (PDE). Based on the CodeBERT model tokenizer, TDE uses natural language processing to measure the semantic distance between test cases. PDE, on the other hand, considers the distance between the source-code paths executed by the test cases. We evaluate these density estimators within EVOSUITE on 100 non-trivial Java classes from the SF110 benchmark. Our results show that the proposed *path-based density estimator* (PDE) outperforms all other density estimators in enhancing mutation scores. It increases mutation scores by 4.26 % on average (with a max of over 60%) to the traditional crowding distance.

Keywords: software testing · search-based software engineering · test case generation · density estimators

1 Introduction

Many-objective evolutionary algorithms (MOEAs) have been used extensively in literature for automatically generating test cases [2,11,20,23]. These algorithms optimize multiple objectives (testing criteria) simultaneously, such as code coverage criteria (*e.g.,* lines, branches) and quality metrics (*e.g.,* mutation score). Previous studies have shown that MOEAs outperform single-objective algorithms in terms of both code coverage and fault detection [6,20,23]. MOEAs have led to various advancements in automated test case generation, such as (1)

G. Jahangirova and F. Khomh (Eds.): SSBSE 2024, LNCS 14767, pp. 18–32, 2024.
https://doi.org/10.1007/978-3-031-64573-0_2

achieving high code coverage [16,23] and (2) having fewer smells [22] compared to manually-written test cases, and (3) detecting unknown bugs [1,15].

One of the open challenges in applying MOEAs to test case generation is maintaining a strong selective pressure [19]. Selective pressure is crucial to guide the search towards the Pareto front, where the best solutions are located. The higher the selective pressure, the more likely the algorithm is to find high-quality solutions [19]. However, maintaining selective pressure is challenging when optimizing a large number of objectives as the search space becomes more complex.

One of the key components in MOEAs that increases the selective pressure is the density estimator. Density estimators are used to measure the density/distributions of solutions in the objective space [18]. These estimators introduce innovative methods for comparing and selecting solutions that would otherwise be non-comparable based solely on dominance criteria [18,19]. The *crowding distance* (CD) is a widely used density estimator in MOEAs [18] and the default density estimator used in NSGA-II [8] and by extension DYNAMOSA [23].

In this paper, we propose two novel density estimators as an alternative to the classical CD, namely the *token-based density estimator* (TDE) and the *path-based density estimator* (PDE). The TDE and PDE are designed to increase the selective pressure within the domain of test case generation as they measure features that are specific to tests. The *token-based density estimator* measures the semantic distance between test cases using the CodeBERT model tokenizer, *i.e.,* if two tests share similar tokens/keywords. We hypothesize that semantically similar test cases are likely to cover similar parts of the code with similar execution states. Conversely, the *path-based density estimator* considers the distance between the different source-code paths executed by the test cases. We hypothesize that test cases that took a different path through the code to reach a node may result in different internal code states.

To evaluate the proposed density estimators, we conducted an empirical study on 100 non-trivial Java classes from the SF110 benchmark [13,23]. We compared the proposed density estimators with two theoretical state-of-the-art density estimators from the evolutionary computation community for many-objective problems, namely the *subvector-dominance assignment* (SD) and the *epsilon-dominance assignment* (ED) [18], and the classical *crowding distance* (CD) [8] *w.r.t.* their ability to generate test suites with higher mutation score, used as a measure for the fault detection capability.

Our results show that the *path-based density estimator* (PDE) outperforms all other density estimators in enhancing mutation scores. It increases mutation scores by 4.26 % on average (with a max of over 60%) to the traditional crowding distance. The classical crowding distance performed the second best in terms of mutation score. The structural coverage of the different density estimators did not show significant differences.

In summary, we make the following contributions:

1. Two novel density estimators designed for automated test case generation.
2. An empirical evaluation of the proposed density estimators on 100 non-trivial Java classes from the SF110 benchmark.

3. A comparison of the proposed density estimators with two state-of-the-art density estimators from the evolutionary computation community and the classical crowding distance.
4. A full replication package containing the results and the analysis scripts [25].

The structure of the paper is as follows: Section 2 provides background information on many-objective test generation and density estimators. Section 3 describes the proposed density estimators while Sect. 4 describes the experimental setup. Section 5 presents the results of the empirical study, and Sect. 6 discusses the threats to validity. Finally, Sect. 7 concludes the paper and outlines future work.

2   Background and Related Work

Previous research has introduced search-based software test generation (SBST) methods that employ meta-heuristics—and genetic algorithms among others— to create tests at various testing levels, including unit [13], integration [9], and system-level testing [3]. Search-based unit test generation is a particularly active area of study in this field, where iterative optimization algorithms evolve tests towards satisfying multiple criteria (*e.g.*, structural coverage, mutation score) for a given class under test (CUT). Prior research indicates that these techniques effectively achieve high code coverage, enhance fault detection [1], and outperform non-SBST-based approaches [16]. Among others, evolutionary algorithms show better performance than large-language models when generating tests for code not available on GitHub [28] (*i.e.*, the training set) and are not impacted by data leakage issues [26]. SBST techniques have proven successful in testing complex systems [17] and for different programming languages [11,13].

Dynamic Many-Objective Sorting Algorithm (DynaMOSA). The state-of-the-art algorithm for unit test generation is a many-objective evolutionary algorithm called DYNAMOSA [23]. Algorithm 1 outlines the pseudo-code of DYNAMOSA [23]. This approach targets multiple coverage elements (*e.g.*, lines, branches, mutants) simultaneously as search objectives. To achieve high scalability, DYNAMOSA utilizes the hierarchy of dependencies between different coverage targets to update the list of objectives dynamically (lines 5 and 10 in Algorithm 1). The list of objectives is updated at each generation by (1) removing already covered targets and (2) adding new targets that are not covered yet but that are structurally depended on the covered ones. This dynamic approach allows to focus the search on the uncovered targets, thus reducing the search space and improving the search efficiency. Recent independent studies [6,20,24] have shown that DYNAMOSA outperforms single-objective and other many-objective evolutionary algorithms *w.r.t.* structural and mutation coverage. Therefore, DYNAMOSA currently is the default algorithm in EVO-SUITE.

Algorithm 1: DynaMOSA

Input:
$U = \{u_1, \ldots, u_m\}$ the set of coverage targets of a program.
Population size M
$G = \langle N, E, s \rangle$: control dependency graph of the program
$\phi : E \rightarrow U$: partial map between edges and targets
Result: A test suite T

1 **begin**
2     $U^* \longleftarrow$ targets in U with not control dependencies
3     $t \longleftarrow 0$ // current generation
4     $P_t \longleftarrow$ RANDOM-POPULATION(M)
5   archive $\longleftarrow$ UPDATE-ARCHIVE$(P_t, \emptyset)$
6     $U^* \longleftarrow$ UPDATE-TARGETS(U^*, G, ϕ)
7   **while** *not (search_budget_consumed)* **do**
8        $Q_t \longleftarrow$ GENERATE-OFFSPRING(P_t)
9   archive $\longleftarrow$ UPDATE-ARCHIVE$(Q_t,$ archive$)$
10       $U^* \longleftarrow$ UPDATE-TARGETS(U^*, G, ϕ)
11       $R_t \longleftarrow P_t \bigcup Q_t$
12       $\mathbb{F} \longleftarrow$ PREFERENCE-SORTING(R_t, U^*)
13       $P_{t+1} \longleftarrow \emptyset$
14       $d \longleftarrow 0$
15   **while** $| P_{t+1} | + | \mathbb{F}_d | \leq M$ **do**
16   CROWDING-DISTANCE-ASSIGNMENT$(\mathbb{F}_d, U^*)$
17         $P_{t+1} \longleftarrow P_{t+1} \bigcup \mathbb{F}_d$
18         $d \longleftarrow d + 1$
19   Sort$(\mathbb{F}_d)$ //according to the crowding distance
20       $P_{t+1} \longleftarrow P_{t+1} \bigcup \mathbb{F}_d[1 : (M - | P_{t+1} |)]$
21       $t \longleftarrow t + 1$
22     $T \longleftarrow$ archive

2.1 Density Estimator for Many-Objective Optimization

A key feature in DYNAMOSA, as in any many-objective algorithms, is the density-estimation method used to increase the selection pressure towards the Pareto front. Selective pressure is achieved by (1) sorting the test cases in the population based on the *preference criterion* (line 12 in Algorithm 1) and (2) selecting the best tests based on their *crowding distance* (line 16 in Algorithm 1). The former promotes the test cases closer to each objective (coverage target), and the latter promotes the test cases more diverse in the objective space. More specifically, the crowding distance is calculated as the sum of the differences in the objective values of the two neighboring test cases [8]. However, as pointed out by Köppen and Yoshida [18], the crowding distance does not scale well with the number of objectives, as it may assign the maximum distance (infinite) to all test cases in the first non-dominated front. To address this limitation, they proposed alternative methods to calculate the crowding distance, such as the *sub-vector dominance* and *epsilon dominance assignment*. We elaborate on these methods in the following sections.

Sub-vector Dominance Assignment. The first alternative estimator introduced by Köppen and Yoshida [18] for many-objective numerical problems is the *sub-vector dominance*. This estimator is applied to each non-dominated front, and therefore, it is used to calculate the solution density for all solutions (test cases in our context) within the same front. The algorithm first assigns an infinite distance to single-member fronts, suggesting that an individual inherently exhibits maximum diversity. For fronts containing multiple solutions, it sets each test case's distance to the highest possible value, indicating that their dominance still needs to be evaluated. Then, this estimator processes each pair of test cases within the same front. For every pair (p1, p2), it assesses their effectiveness against each objective in the set using the corresponding objective values/scores. Two counters `dominate1` and `dominate2` are used to count the number of objectives where p1 performs better than p2 and vice versa. These counters determine how much (the strength) each test case is dominated by the other across all objectives.

After evaluating the dominance of the test cases, the distance for each test case is adjusted. The new distance is the smaller value between its current distance and the number of goals where it is found to be inferior. Essentially, this means that the more a test case is dominated by others, the smaller its distance becomes, reflecting its relative performance deficit across the objectives. Hence, this estimator favors test cases demonstrating superior performance over a broader range of objectives, fostering a varied pool of solutions throughout the evolutionary cycle.

Epsilon-Dominace Assignment. The epsilon-dominance assignment provides a more detailed comparison between solutions than sub-vector dominance. Instead of simply counting how many objectives a solution falls short on, it measures how much worse a solution is in each objective. Similarly, to the other density estimators, this method is applied to each non-dominated front.

For each non-dominated solution p1, this estimators first calculates all ϵ-dominance scores of p1 compared to all other solutions in the same fronts. For each pair of solutions (p1, p2), this metric considers all objective values of p2 that are worse than the corresponding objectives of p1. The epsilon dominance of p1 over p2 is the smallest value ϵ that, if subtracted from all objectives of p2, makes p2 Pareto-dominating p1. This concept is often called additive ϵ-dominance in the related literature [18]. Finally, the density measure of a solution is computed as the smallest of all its epsilon dominance calculated $w.r.t.$ to all other solutions in the same non-dominated front. The larger the distance for a solution p1, the higher the "effort" or the epsilon value needed to make the other solutions dominate p1.

3   Density Estimators for Test Cases

In this section, we present two novel density estimators for test cases as alternative to the crowding distance, i.e., line 16 of Algorithm 1. These estimators are designed to measure (and thus promote) the diversity of the test cases generated by DynaMOSA. The first estimator is based on the semantic content of the test cases (or *genotype*) related to the keywords/tokens that form the test cases. The second estimator works in the objective space (also called *phenotype*) and measures the diversity of the execution paths covered by the test cases. We describe the two estimators in detail and discuss their integration into the main loop of the DYNAMOSA algorithm.

3.1   Token-Based Distance Assignment

We introduce this novel distance assignment with the idea of using natural language processing (NLP) methods to measure the *semantic diversity* of test cases. Our intuition is that at the same level of dominance, test cases that are semantically more diverse should be assigned a higher probability of mating since they may cover different paths in the CUT or use more diverse input values.

To measure the semantic content of the test cases, we rely on the tokenizers of large language models (LLMs), particularly the CodeBERT pre-trained model by Microsoft [12] and publicly available on HuggingFace[1]. The CodeBERT tokenizer is designed to understand both programming and natural languages as it operates similarly to the tokenization process of BERT model [10] but with customization to handle code syntax and semantics.

CodeBERT uses the Byte-Pair Encoding (BPE) algorithm for its tokenization [12], which combines both character- and word-level tokenization. BPE builds an *initial vocabulary* of individual characters and gradually builds up a vocabulary of more frequent and longer sub-word units (byte pairs) by combining pairs of symbols (or characters) that frequently occur together. BPE iteratively counts the frequency of pairs of adjacent symbols in the corpus and merges the most frequent pair to create a new symbol (*iterative learning*). This process is repeated for a predefined number of merge operations, leading to a final vocabulary that includes a mix of characters, common sub-words, and full words. The *tokenization* of new text (test cases in our case) is applied by splitting it into individual characters and then applying the merge rules learned during training. The merging procedure combines characters and sub-words into the tokens present in its final vocabulary.

At the end of the tokenization process, each test case (here considered as text) is tokenized into different tokens, grouped in *special* and *non-special* tokens. The former tokens are essential for the model to understand code structure. For instance, the [SEP] tokens delimit different segments within the input sequence, such as demarcating the end of a code snippet and the beginning of a natural language comment or vice versa. Instead, non-special tokens are the regular tokens representing the input text's content (tests in our case).

Token-Based Density Estimator. Algorithm 2 outlines the pseudo-code of the distance assignment metric based on the token frequency. The algorithm takes in input the current set of objectives U^*, and a list of non-dominated test cases $\mathbb{F}_i$. The algorithm starts by initializing two maps: (1) a mapping of test cases to their respective token sets (TokenMap in line 2), and (2) a mapping of

---

[1] https://huggingface.co/microsoft/codebert-base.

Algorithm 2: Token-based Distance Assignment

Input:
U^*: current set of objectives
$\mathbb{F}_i$ list of non-dominated test cases
Output: Updated test cases with distance values

```
1  begin
2  │    TokenMap ⟵ empty map                          // mapping test cases to tokens
3  │    TokenFrequency ⟵ empty map              // mapping tokens to their frequencies
4  │    foreach τ ∈ 𝔽ᵢ do
5  │    │    text ⟵ TO-TEXT(τ)
6  │    │    tokens ⟵ TOKENIZER(text)                  // Applying CodeBERT tokenizers
7  │    │    TokenMap[τ] ⟵ tokens
8  │    │    /* Update tokens frequencies                                        */
9  │    │    foreach token ∈ tokens do
10 │    │    │    if token ∈ TokenFrequency then
11 │    │    │    │    TokenFrequency[token] ⟵ TokenFrequency[token] + 1
12 │    │    │    else
13 │    │    │    │    TokenFrequency[token] ⟵ 1
14 │    │    │    end
15 │    │    end
16 │    end
17 │    foreach τ ∈ 𝔽ᵢ do
18 │    │    tokens ⟵ TokenMap[τ]
19 │    │    frequency ⟵ ∞
20 │    │    foreach token ∈ tokens do
21 │    │    │    frequency ⟵ MIN(TokenFrequency[token], frequency)
22 │    │    end
23 │    │    SET-DIVERSITY(τ) = 1.0 / frequency
24 │    end
25 end
```

---

tokens to their occurrence frequencies across all test cases (`TokenFrequency` in line 3). Then, the algorithm tokenizes the test and updates the token frequencies among all test cases.

Each test case τ is converted into its list of tokens using the CodeBERT tokenizer (function TOKENIZER in line 6). The resulting tokens are stored in the `TokenMap` and associated with τ in the mapping. Subsequently, the algorithm updates token frequencies stored in `TokenFrequency` by iterating over each token in the tokens set of τ. The token frequencies calculated in lines 9–14 of Algorithm 2 are used to compute a diversity value for each test case with the loop in lines 17–24. Specifically, the algorithm calculates the minimum token frequency for all tokens of a test case τ (lines 19–22). Finally, the assigned distance for τ is calculated as the inverse of this minimum token frequency (line 23).

This token-based metric prioritizes test cases containing rarer tokens, assuming such tests may explore paths or scenarios in the software under test that are less frequently executed. We rely on the CodeBERT tokenizer as it allows us to capture nuances in the code that textual-based methods might miss.

3.2   Path-Based Density Estimator

We proposed a new substitute distance assignment tailored for test case generation and based on dynamic information from the test execution results. Our intuition is that test cases that reach the coverage frontier (i.e., the yet uncov-

---

Algorithm 3: Path-based Distance Assignment

Input:
U^*: current set of objectives
$\mathbb{F}_i$ list of non-dominated test cases
Output: Updated test cases with distance values

```
 1  begin
 2  │   foreach τᵢ ∈ 𝔽ᵢ do
 3  │   │   Lᵢ = COVERED-LINES(τᵢ)                        // set of lines covered by τᵢ
 4  │   │   foreach τⱼ ∈ 𝔽ᵢ do
 5  │   │   │   Lⱼ = COVERED-LINES(τⱼ)                    // set of lines covered by τⱼ
 6  │   │   │   distances(τᵢ, τⱼ) ⟵ JACCARD-DISTANCE(Lᵢ, Lⱼ)
 7  │   │   │   distances(τⱼ, τᵢ) ⟵ distance(τᵢ, τⱼ)
 8  │   │   end
 9  │   end
10  │   visited ⟵ ∅                                     // Set of already-visited test cases
11  │   for index ← 0 to | 𝔽ᵢ | do
12  │   │   bestTest ⟵ ∅                                 // test case to select
13  │   │   maxDiversity ⟵ −∞                            // diversity of the test case to select
14  │   │   foreach τ ∈ 𝔽ᵢ do
15  │   │   │   if τ ∉ visited then
16  │   │   │   │   distance ⟵ AVERAGE-DISTANCE(index, visited, distances)
17  │   │   │   │   /* Select the case with the largest distance to the already considered
                        ones                                                       */
18  │   │   │   │   if distance ⨠ maxDiversity then
19  │   │   │   │   │   maxDiversity ⟵ distance
20  │   │   │   │   │   bestTest ⟵ τ
21  │   │   │   │   end
22  │   │   │   end
23  │   │   end
24  │   │   visited ⟵ visited + {τ}
25  │   │   SET-DIVERSITY(τ) = maxDiversity
26  │   end
27  end
```

---

ered targets/branches) passing through different/diverse execution paths of the software under test are more likely to lead to more diverse execution states (e.g., class attributes and internal variable values).

The new assignment procedure is outlined in Algorithm 3. It leverages line coverage data from previously executed tests in DynaMOSA's early stages, thus avoiding re-execution. For any two test cases, τ_i and τ_j, we compute their *Jaccard distance* based on the sets of lines covered by each, as follows:

$$Jaccard(\tau_i, \tau_j) = 1 - \frac{|\, L_i \cap L_j \,|}{|\, L_i \cup L_j \,|} \tag{1}$$

where L_i and L_j represent the lines covered by τ_i and τ_j, respectively. This metric quantifies the dissimilarity in code coverage between test cases, accounting for all lines covered during execution, including those outside the class under test (e.g., the lines covered for input objects). The pairwise distances are stored in the `distances` matrix in lines 6–7 of Algorithm 3.

Subsequently, our algorithm adopts a greedy strategy to select test cases that maximize diversity iteratively (lines 11–24 in Algorithm 3). Initially, it selects the test case with the highest Jaccard distance from the pre-computed distances matrix `distances`. The selected test is added to the set of visited tests (`visited`) and assigned a distance equal to its maximum Jaccard distance. In each following iteration, the greedy strategy calculates the average Jaccard distance for all test cases that have not been selected yet (if-condition in line 15) using the AVERAGE-DISTANCE function (line 16). This function calculates the average distance of a test case τ to all other previously chosen ones and stored in `visited`. Among the yet-to-select test cases, the algorithm greedily chooses the one with the largest average Jaccard distance (lines 18–21). The diversity of the selected test case (`bestTest`) is then updated to reflect this maximum value (line 25); it is marked as visited (line 24). This process repeats until all test cases are selected and assigned a diversity value, reflecting their contribution to covering diverse execution paths within the software under test.

Code Optimization. To speed up the calculation of the pairwise distances (for our estimators), we pre-allocate a square matrix to store the distances between all test cases, whose dimension (number of columns/rows) is equal to the size of the front. In the worst-case scenario, the font size corresponds to the population size. However, DYNAMOSA can increase the population size if, during the preference criterion calculation, the first front is larger than the population size. In this case, the population size is increased to the first font size, which requires increasing the size of the distance matrix. In case the population size is smaller than the matrix dimension, the latter is not scaled but kept at the largest values in case the population size increases again in subsequent search iterations.

Pre-allocating a matrix of fixed size was critical to (1) speed up the search, as allocating many large matrices incurs a high computational cost, and (2) avoid the overhead of dynamically resizing the matrix during the search. This is also critical to avoid memory exhaustion since creating a new matrix for each iteration will consume more significant memory and at a pace that is too fast for the garbage collector to free the memory. We did experience indeed many memory-related crushes and issues when we did not pre-allocate the matrix.

4   Empirical Study

To investigate the performance of the proposed density estimators within the context of test case generation, we perform an empirical evaluation to answer the following research question:

RQ *How do the proposed density estimators compare to the classical crowding distance w.r.t. mutation score?*

More specifically, we look at the performance of (i) two state-of-the-art density estimators from the evolutionary computation community, namely the *subvector-dominance assignment* (SD) and the *epsilon-dominance assignment* (ED), when applied to the context of test case generation and (ii) two novel density estimators created specifically for test case generation introduced in this work, namely the *token-based density estimator* (TDE) and the *path-based density estimator* (PDE).

4.1   Benchmark

We performed the evaluation on a subset of the SF110 benchmark [14], which is a widely used benchmark in the literature for evaluating test case generation techniques for Java [14,21,23,24]. We do not consider the whole SF110 corpus as many classes are trivial [27] and the total number of classes in the corpus (23,886 Java classes) would take too long to run. Specifically, we randomly selected 100 classes from the SF110 corpus with non-trivial complexity (Cyclomatic Complexity (CC) > 3). This same selection procedure has been used in related literature [21,23,24].

4.2   Parameter Settings

For the parameter settings, we adopted the defaults used by EvoSuite [13] (test case generation tool used in our experiment). These settings have been widely used in literature and previous studies have shown that although parameter tuning impacts the performance of search algorithms, the default parameter values provide reasonable and acceptable results [5]. Therefore, we used DynaMOSA [23] using a single point crossover with a crossover probability of 0.75, mutation with a probability of $1/n$ (n = number of statements in the test case), tournament selection, and a population size of 50. As we are focussing on fault detection, we set *branches* and *strong mutation* as the objectives to optimize. The search budget per unit under test is 300 s.

4.3   Experimental Protocol

To answer the research question, we ran EvoSuite with the four density estimators (SD, ED, TDE, and PDE) and the crowding distance (CD) as a baseline on the 100 classes from the SF110 benchmark and recorded the final branch coverage and mutation score achieved by the generated test cases. To account for the stochastic nature of search-based test case generation, each unit under test was run 20 times. In total, we performed 10 000 runs, consisting of 20 repetitions of 5 configurations on 100 units under test. This required $(10000 \text{ runs} \times 300 \text{ s})/(60 \text{ s} \times 60 \text{ min} \times 24 \text{ h}) \approx 35 \text{ d}$ of consecutive computation time. The experiment was performed on a system with an AMD Ryzen Threadripper PRO 3995WX (64 cores 2.7 GHz) with 256 GB of RAM.

After the experiment, we compared the mutation score achieved by the test cases generated using the different density estimators and performed statistical analysis. We applied the unpaired Wilcoxon signed-rank test [7] with a threshold of 0.05. This non-parametric statistical test determines if two data distributions are significantly different enough to reject the null hypothesis that the two distributions are equal. In addition, we apply the Vargha-Delaney $\hat{A}_{12}$ statistic [29] to determine the effect size of the result, which determines the magnitude of the difference between the two data distributions.

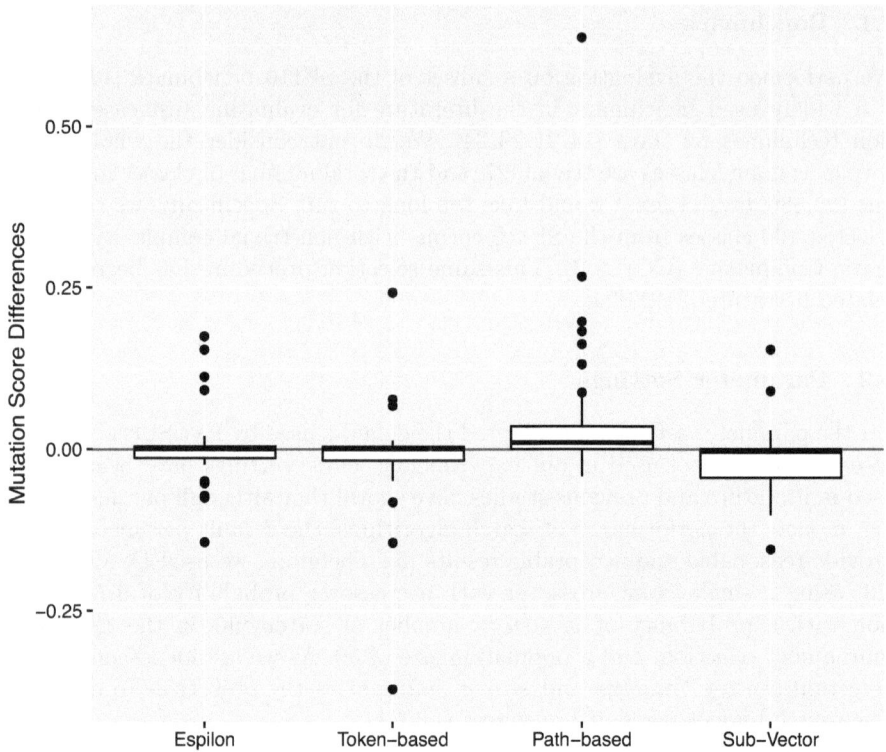

Fig. 1. Difference in achieved mutation score across the classes in the benchmark using the different density estimators compared to the crowding distance

5   Results

This section presents the results of our empirical study. All differences in results are presented in absolute differences (percentage points).

Figure 1 shows the difference in the mutation score achieved by the test cases generated using the different density estimators compared to the classical crowding distance. The datapoints in the boxplot represent the difference in the median mutation score for each class in the benchmark. The results show that the *path-based density estimator* (PDE) achieves the highest mean mutation score (53.90 %) across the classes in the benchmark and improves the most over crowding distance. The crowding distance (CD) achieves a mean mutation score of 49.64 %. The *token-based density estimator* (TDE) has a mean mutation score of 49.17 %, which is slightly lower than the crowding distance. The *epsilon-dominance assignment* (ED) and the *subvector-dominance assignment* (SD) achieve mean mutation scores of 48.83 % and 48.64 %, respectively.

We, additionally, performed a statistical analysis to determine the significance of the differences in the mutation score achieved by the test cases generated using the different density estimators. Table 1 shows the results of this

Table 1. Results of the statistical analysis of the achieved mutation score using Vargha-Delaney $\hat{A}_{12}$ statistic

Comparison	#Win			#No diff.	#Lose		
	Large	Medium	Small	Negl	Small	Medium	Large
Path-based vs. Token-based	18	11	4	63	-	2	2
Path-based vs. Epsilon	16	10	7	62	1	3	1
Path-based vs. Sub-Vector	17	12	2	66	-	3	-
Path-based vs. Crowding	12	12	3	71	1	1	-
Token-based vs. Epsilon	2	7	3	73	3	8	4
Token-based vs. Sub-Vector	7	8	-	76	1	5	3
Token-based vs. Crowding	2	2	-	85	2	4	5
Epsilon vs. Sub-Vector	3	2	2	90	-	2	1
Epsilon vs. Crowding	4	1	4	79	3	6	3
Sub-Vector vs. Crowding	3	2	-	82	1	5	7

statistical analysis based on a p-value ≤ 0.05. In this table, the *#Win* columns indicate the number of times that the left density estimator has a statistically significant improvement over the right one, the *#No diff.* column indicates the number of times that there is no evidence that the two competing density estimators are different, and the *#Lose* columns indicate the number of times that the left density estimator has statistically worse results than the right one. The *#Win* and *#Lose* columns also include the $\hat{A}_{12}$ effect size, classified into *Small*, *Medium*, and *Large*.

The results show that the *path-based density estimator* (PDE) outperforms the other density estimators in most comparisons. In particular, PDE outperforms the *epsilon-dominance assignment* (ED) in 33 out of 100 comparisons, the *token-based density estimator* (TDE) in 33 out of 100 comparisons, the *subvector-dominance assignment* (SD) in 31 out of 100 comparisons, and the *crowding distance* (CD) in 27 out of 100 comparisons. The *epsilon-dominance assignment* (ED) outperforms the *token-based density estimator* (TDE) in 15 out of 100 comparisons, the *subvector-dominance assignment* (SD) in 7 out of 100 comparisons, and the *crowding distance* (CD) in 9 out of 100 comparisons. The *token-based density estimator* (TDE) outperforms the *subvector-dominance assignment* (Sub-Vector) in 15 out of 100 comparisons and the *crowding distance* (CD) in 4 out of 100 comparisons. Lastly, the *subvector-dominance assignment* (SD) outperforms the *crowding distance* (CD) in 5 out of 100 comparisons. Interestingly, the classical crowding distance (CD) performs better in more cases than the *subvector-dominance assignment* (SD), the *epsilon-dominance assignment* (ED), and the *token-based density estimator* (TDE). However, in the majority of the classes in the benchmarks, there is no significant difference between the density estimators.

In addition to the mutation score, we also looked at the branch coverage achieved by the generated test cases. We observed that the branch coverage achieved by the test cases generated using the different density estimators is identical. This indicates that the difference in mutation score is not due to differences in branch coverage but rather due to the improvement in the density estimators.

6   Threats to Validity

This section discusses the potential threats to the validity of our study.

External Validity: One of the threats to the *external validity* of our study is the selection of the benchmark. The selection of the benchmark impacts the generalizability of the results. To address this threat, we used a subset of the SF110 benchmark, which is a widely used benchmark in the literature for evaluating test case generation techniques for Java. The subset of the benchmark was selected based on the complexity of the classes to ensure that the results are not biased by trivial classes. However, the results may not generalize to other benchmarks or programming languages.

Conclusion Validity: The stochastic nature of search-based test case generation introduces a threat to the *conclusion validity* of our study. To mitigate this threat, we ran each configuration 20 times with different random seeds. This allows us to draw statistically significant conclusions from the results. We have followed the best practices for running experiments with randomized algorithms as laid out in well-established guidelines [4]. Additionally, we used the unpaired Wilcoxon signed-rank test and the Vargha-Delaney $\hat{A}_{12}$ effect size to assess the significance and magnitude of our results.

7   Conclusions and Future Work

In this paper, we have presented two novel density estimators for automated test case generation to increase the selective pressure within the search front. We compared the proposed density estimators with two state-of-the-art density estimators from the evolutionary computation community and the classical crowding distance. Our results show that our proposed *path-based density estimator* (PDE) is the most effective in promoting the diversity of the solutions in the population, leading to a better spread of the solutions in the objective space and a higher mutation score. The classical crowding distance performed the second best in terms of mutation score.

In future work, we will evaluate the proposed density estimators on other test generation problem—*e.g.*, system-level test case generation—and other software testing problems, diversity-based test case prioritization. We also plan to (1) use different tokenizers as well as (2) different LLMs for the test case embeddings as alternatives to CodeBERT. Finally, we plan to analyze the relation between the diversity of the test cases and the fault detection capability of the generated test suites.

References

1. Almasi, M.M., Hemmati, H., Fraser, G., Arcuri, A., Benefelds, J.: An industrial evaluation of unit test generation: finding real faults in a financial application. In: 2017 IEEE/ACM 39th International Conference on Software Engineering: Software Engineering in Practice Track (ICSE-SEIP), May 2017, pp. 263–272. IEEE (2017). https://doi.org/10.1109/ICSE-SEIP.2017.27
2. Arcuri, A.: Test suite generation with the many independent objective (MIO) algorithm. Inf. Softw. Technol. **104**, 195–206 (2018)
3. Arcuri, A.: RESTful API automated test case generation with EvoMaster. ACM Trans. Softw. Eng. Methodol. (TOSEM) **28**(1), 1–37 (2019)
4. Arcuri, A., Briand, L.: A hitchhiker's guide to statistical tests for assessing randomized algorithms in software engineering. Softw. Test. Verification Reliab. **24**(3), 219–250 (2014)
5. Arcuri, A., Fraser, G.: Parameter tuning or default values? An empirical investigation in search-based software engineering. Empir. Softw. Eng. **18**, 594–623 (2013)
6. Campos, J., Ge, Y., Albunian, N., Fraser, G., Eler, M., Arcuri, A.: An empirical evaluation of evolutionary algorithms for unit test suite generation. Inf. Softw. Technol. **104**, 207–235 (2018). https://doi.org/10.1016/j.infsof.2018.08.010
7. Conover, W.J.: Practical Nonparametric Statistics, vol. 350. Wiley, Hoboken (1999)
8. Deb, K., Pratap, A., Agarwal, S., Meyarivan, T.: A fast and elitist multiobjective genetic algorithm: NSGA-II. IEEE Trans. Evol. Comput. **6**(2), 182–197 (2002)
9. Derakhshanfar, P., Devroey, X., Panichella, A., Zaidman, A., van Deursen, A.: Towards integration-level test case generation using call site information. arXiv preprint arXiv:2001.04221 (2020)
10. Devlin, J., Chang, M.W., Lee, K., Toutanova, K.: BERT: pre-training of deep bidirectional transformers for language understanding. arXiv preprint arXiv:1810.04805 (2018)
11. Erni, N., Mohammed, A.A.M.A., Birchler, C., Derakhshanfar, P., Lukasczyk, S., Panichella, S.: SBFT tool competition 2024–Python test case generation track. arXiv preprint arXiv:2401.15189 (2024)
12. Feng, Z., et al.: CodeBERT: a pre-trained model for programming and natural languages. arXiv preprint arXiv:2002.08155 (2020)
13. Fraser, G., Arcuri, A.: EvoSuite: automatic test suite generation for object-oriented software. In: Proceedings of the 19th ACM SIGSOFT Symposium and the 13th European Conference on Foundations of Software Engineering, pp. 416–419 (2011)
14. Fraser, G., Arcuri, A.: A large-scale evaluation of automated unit test generation using EvoSuite. ACM Trans. Softw. Eng. Methodol. (TOSEM) **24**(2), 1–42 (2014)
15. Fraser, G., Arcuri, A.: 1600 faults in 100 projects: automatically finding faults while achieving high coverage with EvoSuite. Empir. Softw. Eng. **20**, 611–639 (2015)
16. Jahangirova, G., Terragni, V.: SBFT tool competition 2023-Java test case generation track. In: 2023 IEEE/ACM International Workshop on Search-Based and Fuzz Testing (SBFT), pp. 61–64. IEEE (2023)
17. Khatiri, S., Saurabh, P., Zimmermann, T., Munasinghe, C., Birchler, C., Panichella, S.: SBFT tool competition 2024: CPS-UAV test case generation track. In: 17th International Workshop on Search-Based and Fuzz Testing (SBFT), Lisbon, Portugal, 14–20 April 2024. ZHAW Zürcher Hochschule für Angewandte Wissenschaften (2024)
18. Köppen, M., Yoshida, K.: Substitute distance assignments in NSGA-II for handling many-objective optimization problems. In: Obayashi, S., Deb, K., Poloni,

C., Hiroyasu, T., Murata, T. (eds.) EMO 2007. LNCS, vol. 4403, pp. 727–741. Springer, Heidelberg (2007). https://doi.org/10.1007/978-3-540-70928-2_55

19. Li, B., Li, J., Tang, K., Yao, X.: Many-objective evolutionary algorithms: a survey. ACM Comput. Surv. (CSUR) **48**(1), 1–35 (2015)

20. Lukasczyk, S., Kroiß, F., Fraser, G.: An empirical study of automated unit test generation for Python. Empir. Softw. Eng. **28**(2), 36 (2023)

21. Molina, U.R., Kifetew, F., Panichella, A.: Java unit testing tool competition: sixth round. In: Proceedings of the 11th International Workshop on Search-Based Software Testing, pp. 22–29 (2018)

22. Panichella, A., Panichella, S., Fraser, G., Sawant, A.A., Hellendoorn, V.: Test smells 20 years later: detectability, validity, and reliability. Empir. Softw. Eng. **27**, 170 (2022). https://doi.org/10.1007/s10664-022-10207-5

23. Panichella, A., Kifetew, F.M., Tonella, P.: Automated test case generation as a many-objective optimisation problem with dynamic selection of the targets. IEEE Trans. Softw. Eng. **44**(2), 122–158 (2017)

24. Panichella, A., Kifetew, F.M., Tonella, P.: A large scale empirical comparison of state-of-the-art search-based test case generators. Inf. Softw. Technol. **104**, 236–256 (2018)

25. Panichella, A., Mitchell, O.: Replication package of "higher fault detection through novel density estimators in unit test generation", May 2024. https://doi.org/10.5281/zenodo.11209898

26. Sallou, J., Durieux, T., Panichella, A.: Breaking the silence: the threats of using LLMs in software engineering. In: ACM/IEEE 46th International Conference on Software Engineering. ACM/IEEE (2024)

27. Shamshiri, S., Rojas, J.M., Fraser, G., McMinn, P.: Random or genetic algorithm search for object-oriented test suite generation? In: Proceedings of the 2015 Annual Conference on Genetic and Evolutionary Computation, pp. 1367–1374 (2015)

28. Siddiq, M.L., Santos, J., Tanvir, R.H., Ulfat, N., Rifat, F.A., Lopes, V.C.: Exploring the effectiveness of large language models in generating unit tests. arXiv preprint arXiv:2305.00418 (2023)

29. Vargha, A., Delaney, H.D.: A critique and improvement of the CL common language effect size statistics of McGraw and Wong. J. Educ. Behav. Stat. **25**(2), 101–132 (2000)

Many Independent Objective Estimation of Distribution Search for Android Testing

Michael Auer[✉], Andreas Strobl, and Gordon Fraser

University of Passau, Passau, Germany
{M.Auer,Gordon.Fraser}@uni-passau.de

Abstract. Search-based test generation techniques have shown promising results when applied to Android. The most common approach is to use genetic algorithms, but defining adequate crossover and mutation operators is challenging for Android testing: The actions that form tests are often state-dependent, which implies that they cannot be arbitrarily re-arranged without leading to non-executable tests. In this paper, we therefore investigate the use of estimation of distribution algorithms (EDAs), which are search algorithms where probability distributions over the input space are adapted and sampled instead of using explicit variation operators. We introduce MIOEDA, a many-objective search algorithm that integrates the Many Objective Independent (MIO) search algorithm, which was specifically designed for test generation, with estimation of distribution search, thus enabling the search for code coverage without requiring classical variation operators. Using our implementation of MIOEDA as part of the open source Android test generator MATE for an evaluation study on 100 Android apps demonstrates that MIOEDA can serve as a successful replacement of search algorithms based on traditional variation operators.

Keywords: Android · MIO · EDA · Automated Test Generation

1 Introduction

Various test generation approaches have been explored to reduce the human effort involved in testing Android apps. Search-based testing, primarily through the form of genetic algorithms [3,14,15], has shown promising results. However, one particular difficulty that arises when applying genetic algorithms to Android is the creation of eligible crossover and mutation operators: Since the individual actions that form a test case are often state dependent, they cannot be arbitrarily re-arranged during crossover without breaking the test. Likewise, changing, inserting or deleting individual actions as part of mutation may break test sequences. Existing approaches try to circumvent this problem by defining complex variation operators that take these dependencies into account [2,12,18], avoid dependencies by interacting with screen coordinates rather than widgets [13], or by applying crossover at the test suite level [13].

© The Author(s), under exclusive license to Springer Nature Switzerland AG 2024
G. Jahangirova and F. Khomh (Eds.): SSBSE 2024, LNCS 14767, pp. 33–48, 2024.
https://doi.org/10.1007/978-3-031-64573-0_3

While such variation operators are an integral part of many metaheuristic search algorithms, the family of estimation of distribution algorithms (EDAs) [9] stands out in that it can operate without explicit variation operators, and instead samples individuals from a probabilistic model, which is iteratively refined with the information gained through fitness evaluations [11]. Although EDAs have been successfully applied in many different domains [9], their use in the context of Android test generation has not seen much attention so far.

In this paper we therefore investigate a novel search-based test generation algorithm that combines the benefits originating from EDAs (requiring no variation operators) with current state-of-the-art search-based test generation. In particular, our approach builds on top of the Many Independent Objective (MIO) algorithm [3], which was designed specifically to deal with the large number of coverage objectives during test generation. The MIO algorithm uses no crossover, and instead of mutating individuals we use a probabilistic model that encodes information about each objective in a tree-like structure [17]. A new test can then be directly sampled from the model and the iterative refinement ensures that the sampled tests are optimised towards satisfying the individual objectives without the need to perform exploitation through mutation.

We implemented MIOEDA as an extension of the MATE [7] search-based test generator for Android, and empirically evaluate it on a set of 100 F-Droid apps. The results indicate that MIOEDA provides a viable alternative for the test generation in Android. In particular, MIOEDA is slightly ahead of the basic MIO approach using classical mutation, a random exploration baseline, and the related approach of the state-of-the-art tool STOAT [19].

2   Background

2.1   Automated Android Testing

Search-based algorithms [2,12,13,18] are a popular variant among the different approaches to generate tests for Android [10]. In particular, genetic algorithms evolve a population of chromosomes (i.e., tests) with respect to one or multiple objective functions, such as branch coverage. As part of the evolution, the individuals undergo crossover and mutation to form new chromosomes for the next generation. However, the construction of those search operators for Android is difficult because changes to a system test may result in the test no longer being executable if state dependencies are violated.

One approach to overcome this problem is to design the crossover and mutation operators such that they operate only on individual segments [12], i.e., parts of the app under test (AUT) that can be traversed independently, e.g., activities. This can also be achieved by leveraging a GUI model to derive valid offspring during crossover [18]. In particular, the cutpoint between two tests can be chosen such that the combined action sequence remains executable according to the model after crossover. To avoid handling dependencies explicitly, other proposed approaches include repeating crossover until a valid offspring is formed [2], or defining tests as sequences of actions that are state independent, i.e., they are not coupled to a specific widget but solely on its coordinates [13]. This allows to arbitrarily reassemble actions while maintaining an executable test, although certain actions might have no effect, e.g., a click on a coordinate that is not backed up by a clickable widget. Finally, when evolving test suites rather than test cases, crossover only re-arranges tests rather than modifying them [8,13]. However, mutation nevertheless requires modifying individual tests.

The Stoat [19] approach is an exception in that it uses search, but no explicit variation operators. In a first phase, Stoat tries to explore the AUT as much as possible using a dynamic exploration strategy and thereby constructs a GUI model that records which actions have been executed how often in a particular state. In a second phase, this GUI model is converted to a probabilistic model in which the initial action probabilities are derived from the recorded execution frequencies as well as the action type, e.g., clicks are favoured over scrolling actions. The action probabilities are mutated, and test suites sampled from the model using Gibbs sampling [19] are evaluated using a linear combination of model coverage, code coverage and test diversity. This is similar to our objective of using an EDA, but does not provide the ability to optimise for individual objectives (e.g., branches, activities, etc.), which in other testing domains is done using many-objective optimisation algorithms (MOAs) [3,14,15].

2.2   Estimation of Distribution Algorithms

Estimation of distribution algorithms are evolutionary algorithms, but in contrast to genetic algorithms, EDAs can operate without crossover and mutation operators [11]. Individuals are sampled from a probabilistic model that describes the input space, their fitness is evaluated, and a selection operator chooses the most promising candidates, which in turn are used to update the probabilistic model [9]. There are different variants of EDAs [9], and one distinguishing factor is the representation of the probabilistic model. In the simplest case this can be a probability vector, but more complex input structures require more complex models. Since in Android testing there are dependencies between actions (i.e., actions are state dependent), the parameters of actions are most often real-valued, and the number of actions per test case may be variable, tree-like structures are more appropriate. A popular such variant is used in the *Probabilistic Incremental Program Evolution (PIPE)* [17] algorithm, which was originally applied to derive programs consisting of sequences of instructions. This can be mapped to our context, where programs represent test cases and the instructions refer to the individual actions. PIPE encodes the probabilistic model in a *probabilistic prototype tree* (PPT) from which new programs are sampled by traversing the tree from the root to a terminal node. As these paths can be of varying length, they ideally fit the structure of a test case that is also composed of a variable number of actions. Moreover, the state dependency of actions can be adequately mapped by the parent-child relation (edge connection) of two nodes, i.e., an action leading to the current state (child node) is only applicable if the predecessor state (parent node) has been previously traversed.

2.3   Many Independent Objective Search for Code Coverage

Test generation with the aim to achieve high code coverage is a many-objective problem where objectives usually refer to coverage goals (e.g., branches). Popular algorithms include the Many Objective Sorting Algorithm (MOSA [14,15]) and Many Independent Objective (MIO) algorithm [3]. The latter is of particular relevance for our context, since it is designed for problems where the number of objectives can become exceedingly large. The MIO algorithm (see Algorithm 1) maintains an archive for each objective k (e.g., branch), in which a population of up to n tests is stored. Initially the archive is empty (Algorithm 1, line 1) and the first test that is generated randomly is inserted (line 2–3).

From this step onwards, MIO either samples a test randomly (line 5–6) or from the archive (line 7–8) until the termination condition is met. Tests sampled from the archive are mutated (line 9), where the parameter m controls how many mutations should be performed. Then, the fitness is evaluated with respect to each objective and the archive is updated (line 11–14). If a test covers one or more objectives, any previously archived tests for those objectives are replaced by the new test. If the archive already stores a test that covers an objective, the test is only replaced if it is better on a secondary criterion, e.g., length. If the test does not cover an objective and the archive does not already include a test that covers the objective, the test is inserted in the archive under the condition that the population for the respective objective is not yet full. Otherwise, the test replaces the least favorable test in the population assuming that the test is better. A test that gets assigned the worst possible fitness value is ignored.

At the end of each generation P_r, n and m are updated (line 15) depending on the parameter F denoting the beginning of the focused search phase, in which more effort is put towards exploitation rather than exploration. This means that the values of the parameters P_r and n decrease while the mutation rate m linearly increases over time until the focused search begins, at which point P_r is set to 0 (tests are no longer sampled randomly but solely from the archive).

3   Combining MIO with EDA

We explore the use of EDAs for Android test generation by adapting and integrating the PIPE algorithm (cf. Subsect. 2.2) into the state-of-the-art MIO search algorithm. This requires the encoding of the probabilistic model through an Application State Tree (AST) (cf. Subsect. 3.1) which is sampled for test cases using an adaptation of PIPE's original sampling procedure (cf. Subsect. 3.2). After a new population has been drawn, the fitness of each test is evaluated and the probabilistic model is refined (cf. Subsect. 3.3). Since MIO optimises towards multiple objectives, a dedicated probabilistic model is used for each objective, which replaces the archive (cf. Subsect. 3.4).

3.1   Application State Tree

The probabilistic model used in our approach is an Application State Tree (AST), which differs from PIPE's PPT in that the nodes refer to states of the AUT and the actions applicable in a given state form the connections to adjacent states. More formally, an AST $t = (N, E)$, consists of $|N|$ nodes interconnected by $|E|$ edges. Each node $n = (s, d) \in N$ contains a state s and a probability distribution d over the applicable actions in that state. An edge $e = ((s_1, d_1), a, (s_2, d_2)) \in E$ describes a transition from node $n_1 = (s_1, d_1)$ to node $n_2 = (s_2, d_2)$, triggered by action a in state s_1. In our context, a state s represents the UI component hierarchy. We adopt an abstraction to prevent the explosion of the search space. Specifically, two states are considered equal if they satisfy the following conditions: (1) they belong to the same activity, (2) originate from the same package, and (3) possess identical widgets. Widget equality is determined by two factors: (1) identical screen positions and (2) matching height and width dimensions. An AST encodes the state sequences encountered by all tests. For instance, the tests and their state sequences in Fig. 1a can be represented as shown in Fig. 1b. Note that a transition into the same state is represented by two distinct nodes, e.g., *S2* to *S2* of the first test in Fig. 1a.

Algorithm 1. Many Independent Objective Algorithm

Require: $k > 0$ ▷ Number of objectives.
Require: $n > 0$ ▷ Maximum population size for each objective in the archive.
Require: $P_r \in [0, 1]$ ▷ Probability for sampling a test randomly.
Require: $F \in [0, 1]$ ▷ Percentage of the search budget when focused search starts.
Require: $m > 0$ ▷ Mutation rate.
 1: $T_{k,n} \leftarrow emptyArchive()$ ▷ Maintains for each objective k a population $|T_k| <= n$.
 2: $test \leftarrow randomSample()$
 3: $updateArchive(T_{k,n}, test)$
 4: **while** $terminationConditionIsNotMet()$ **do**
 5:     **if** $P_r > rand()$ **then**
 6:         $test \leftarrow randomSample()$
 7:     **else**
 8:         $test \leftarrow archiveSample(T_{k,n})$
 9:         $test \leftarrow mutate(test, m)$
10:     **end if**
11:     **for** $i = 0$ to k **do**
12:         $fitness \leftarrow getFitness(test, i)$
13:         $updateArchive(T_{k,n}, test, fitness, i)$
14:     **end for**
15:     $updateParameters(F, P_r, n, m)$
16: **end while**

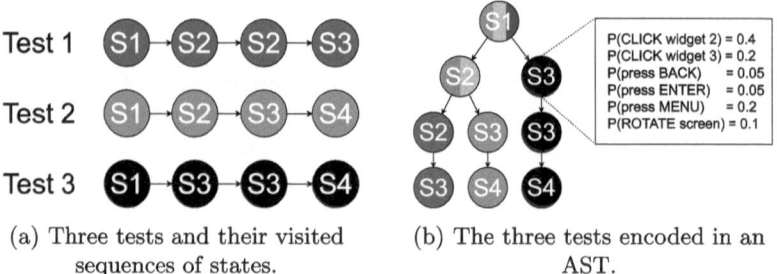

(a) Three tests and their visited
sequences of states.

(b) The three tests encoded in an
AST.

Fig. 1. Tests and their encoding as AST.

3.2 Sampling Tests from the AST

To adapt PIPE for Android testing, we modify the sampling procedure: Rather than sampling a complete tree from a probabilistic prototype tree (PPT), we sample only sequences from the root node of an AST down to a leaf node. This path represents a sequence of states $s_1, s_2, \ldots, s_n$ connected by actions $a_1, a_2, \ldots, a_{n-1}$, constituting a single test. The sequence is constructed dynamically since the outcome after applying an action a_i cannot be predetermined. Thus, iterative model updates are necessary, maintaining the current position $n_m = (s_m, d_m) \in N$. When selecting a new action, we probabilistically sample an action a_m from the distribution d_m of the current state s_m. Upon executing a_m, we either arrive at a known state s_k, corresponding to an existing node n_k, or we encounter a new state. In the former scenario, we simply update the current position to $n_m = n_k$. In the latter, we create a new node $n = (s, d)$, initialising the probability distribution d for available actions in state s and updating the current position to $n_m = n$. The initial action probabilities are set based on Stoat's heuristics [19]. Subsequently, we verify whether the last action led to a crash or transitioned to a state outside the app. If so, the test is returned; otherwise, the loop continues until a predefined number of actions is reached.

3.3 Probability Updates

After a complete population has been sampled, the probabilistic model is updated as outlined in Algorithm 2. First, the best test case from the current population is selected (line 1). Then, this test becomes the elite if it is better than the current one, i.e., the globally best test seen so far (line 2–4). Next, elitist learning is performed in a loop with a certain probability (line 5–7), followed by generation-based learning (line 8) and a possible mutation step (line 9–11).

Both elitist and generation-based learning invoke the same procedure but with a different argument. The objective here is to adjust the action probabilities such that the elite or current best test is likely redrawn. We start by mapping the action sequence of the test to the corresponding path in the AST. Then, we split this path into two sub paths by determining the point where an action or

Algorithm 2. PIPE model update

1: $currentBest \leftarrow testWithHighestFitness(population)$
2: **if** $fitness(currentBest) < fitness(elite)$ **then**
3:    $elite \leftarrow currentBest$
4: **end if**
5: **while** $random() < P_{el}$ **do**
6:    $adaptTreeTowards(elite)$
7: **end while**
8: $adaptTreeTowards(currentBest)$
9: **if** $random() < P_{mutate}$ **then**
10:    $treeMutation(currentBest)$
11: **end if**

any subsequent action of the test no longer increases the fitness. We denote the former sequence as the path of good nodes, while the latter sequence refers to the path of bad nodes, respectively. Then, we start rewarding the actions of the test that are associated with the good nodes. We determine the current path probability of the good nodes by multiplying the individual probabilities along that path. After that, we compute a new target probability for this path based on the current one as described by the following formula:

$$betterProb(currentProb, fitness) =$$

$$currentProb + (1 - currentProb) * lr * \frac{\epsilon + fitness_{elite}}{\epsilon + fitness}$$

The constant lr denotes the positive learning rate and controls by how much the current probability should be increased depending on the fitness of the elite test seen so far and the current test. We continually iterate over the good nodes and increment the probability of the corresponding test's action as follows:

$$newProb(probBefore) = \quad probBefore + c^{lr} * lr * (1 - probBefore)$$

The new probability is derived from the previous probability, the positive learning rate lr and a constant c^{lr}. After a complete traversal over the good nodes, the path probability is recomputed. If the target probability is reached, the rewarding procedure stops, otherwise a further traversal over the good nodes is performed. Similarly, a new target probability is computed over the path of bad nodes as shown by the following formula:

$$worseProb(currentProb, fitness) =$$

$$currentProb - currentProb * nlr * \frac{\epsilon + fitness}{\epsilon + fitness_{elite}}$$

The constant nlr denotes the negative learning rate and controls by how much the current probability should be decreased. Then, the probabilities of the test's actions associated with the bad nodes are decreased continually as follows:

$$newProb(probBefore) = probBefore - c^{lr} * nlr * probBefore$$

Since we modified the probabilities of certain actions along the path of good and bad nodes, we need to normalise the probabilities to ensure a valid probability distribution at each node [17].

Finally, with a probability of P_{mutate}, PIPE applies a mutation step (line 9–11) that encourages exploration around the best solution [17]. It traverses the tree along the nodes of the current best test case and adapts the probabilities of available actions with a likelihood of $\frac{P_{mutate}}{\sqrt{|test|}}$ according to the following formula:

$$newProb(probBefore) = probBefore + mr * (1 - probBefore)$$

Here, mr denotes the mutation rate and controls by how much the probability should be increased. Decreasing the action probability at this stage would have a contradicting effect, since the likelihood of selecting a different action in an early state would possibly lead to a completely different test. Increasing the probability shifts this effect towards the end, thus leading to a test that is similar but diverse.

3.4 Integrating PIPE and MIO

In order to integrate the PIPE model with the MIO algorithm (cf. Algorithm 1), we need to replace the archive (line 1) with an instance of the PIPE model for each objective. When MIO would originally sample a test from the archive and mutate it (line 8–9), it now samples a test from the probabilistic model associated with the lowest sampling counter. If multiple models share an identical value, a random selection is made. The counter is increased afterwards and reset to zero in case the test improved fitness [3]. This promotes the selection of a model that has not been considered recently but at the same time is likely coverable according to the associated objective function. Instead of updating the archive (line 3 & 13), we evaluate the fitness of the sampled test with respect to each objective and update the probabilistic models separately (cf. Algorithm 2). Similar to MIO, the parameters that control the balance between exploration and exploitation (line 15) are updated after each generation, but only with respect to P_r, since MIOEDA does not perform mutation on sampled tests.

4 Evaluation

We aim to answer the following research question:

RQ: How does MIOEDA compare to random exploration and the state-of-the-art search-based approaches MIO and STOAT in terms of coverage?

4.1 Implementation

We implemented the proposed MIOEDA algorithm in the open source Android test generator MATE [7]. All experiments were conducted on a compute cluster, where each node is equipped with two Intel Xeon E5-2620v4 CPUs (16 cores) with 2.10 GHz and 256 GB of RAM, and runs Debian GNU/Linux 11 with Java 11. We limited executions of MATE to four cores and 60 GB of RAM, where the emulator (Pixel XL) runs a x86 image with API level 25 (Android 7.1.1) and is limited to 4 GB of RAM with a heap size of 576 MB.

4.2   Parameter Tuning

To determine appropriate values for the core parameters of MIOEDA we performed a tuning study on a dataset consisting of 10 randomly sampled *F-Droid* apps. We evaluated each parameter configuration ten times each lasting 1h to compensate effects caused by random noise. The tested parameters together with their possible values and the finally best value per parameter are listed in Table 1. The possible values were chosen *around* the default value (highlighted in bold) originating either from the introductory paper of the MIO [3] or PIPE [17] algorithm, respectively, as well as based on preliminary results. Since the time required to perform a complete grid search comprising all parameters was not feasible, we first tested each parameter independently. Then, we further evaluated configurations where a combination of parameters likely has a coupling effect, e.g., the two MIO parameters P_r and F. In total, we tested 33 parameter configurations. To determine whether one configuration is better than one or more other configurations, we applied the following tournament ranking: For each pair of configurations (c_1, c_2) we compare the two configurations for each of the apps in terms of activity and branch coverage using a Wilcoxon-Mann-Whitney U test with a 95% confidence level. If a statistical difference is observed in any coverage metric, we use the Vargha-Delaney $\hat{A}_{12}$ effect size [20] to determine which of the two configurations is better, and the *score* for this configuration and the particular coverage metric is increased by one. At the end, the configuration with the highest combined score (sum) is the overall best configuration.

Table 1. The fine-tuned parameters used in the MIOEDA algorithm; default values highlighted in bold.

Parameter	Possible Values	Best Value
Probability random sampling P_r	0.0/0.2/**0.5**/0.8	0.0
Focused search start F	0.2/**0.5**/0.8	0.5
Learning rate lr	**0.01**/0.05/0.2/0.5	0.2
Negative learning rate nlr	0.01/0.05/**0.2**/0.5	0.2
Elitist learning probability elp	0.0/**0.1**/0.3/0.7	0.3
Mutation probability mp	0.0/0.2/**0.4**/1.0	0.4
Mutation rate mr	0.2/**0.4**/1.0	0.4

We performed the same tuning for the four core parameters of the MIO algorithm. From the 19 configurations we tested the configuration ($P_r = 0.0$, $F = 0.8$, $n = 10$, $m = 5$) yielded the best results. We refer the interested reader to the replication package[1] that contains a detailed description of the tested parameter configurations including their individual tournament ranking scores.

---

[1] https://doi.org/10.6084/m9.figshare.25556967.

4.3   Study Subjects

We randomly sampled a data set consisting of 100 apps from *F-Droid*[2]. During sampling we excluded apps that match one of the following criteria: (i) The app represents a game; (ii) the APK is not suitable for the x86 architecture or API level 25 to allow running all apps on the same emulator; (iii) the APK could not be re-built after being instrumented; (iv) the app crashed/behaved unexpected after being re-signed; (v) MATE has troubles to interact with the app; (vi) the app appeared already in the tuning dataset.

4.4   Experiment Procedure

We compare MIOEDA with MIO and a baseline random exploration strategy implemented in MATE as well as the STOAT tool using the 100 study subjects. We repeated each 3h run lasting 15 times to reduce random influences and recorded activity as well as branch coverage. To determine whether one algorithm performs better than the other, we use the Wilcoxon-Mann-Whitney U test and the Vargha-Delaney $\hat{A}_{12}$ effect size [20] metric. We configured STOAT to use 60 min for model construction (phase 1) and 120 min for stochastic sampling (phase 2) [19]. MIOEDA and MIO were both configured to optimise towards branches with the commonly-used approach level and branch distance fitness function [3]; the remaining parameters were chosen based on the results of the tuning study. MIO was further configured to use the default mutation operator of MATE [18], where essentially a random cut point is chosen from the test case action sequence and the actions from the cut point onwards are replaced with arbitrary actions applicable in the respective state.

4.5   Threats to Validity

Threats to external validity may arise from our sample of subject apps, and results may not generalise beyond the tested apps. To counteract selection bias, we picked the 100 apps for the empirical study randomly. There may also be some implicit bias, e.g., the apps on *F-Droid* might be simpler than those on *Google PlayStore*. We also stuck to one specific emulator configuration and API level, but results may differ on other versions. *Threats to internal validity* may arise from bugs in MATE or our analysis scripts. To mitigate this risk, we manually reviewed the results, and tested and reviewed all code. To reduce the risk of favouring one algorithm over the other, we used the same default parameters wherever applicable, e.g., the maximum number of actions per test case was fixed to 50 actions. In addition, MIOEDA, MIO and the random exploration strategy were all implemented in the same test generator, thus lowering the risk of favouring one tool's implementation. To make coverage results comparable, we instrumented all apps with the same coverage tool and adapted STOAT to use the same coverage mechanism than MATE, as done previously [6]. *Threats*

---

[2] https://f-droid.org/.

to construct validity may result from our choice of metrics, in particular activity and branch coverage. However, these are standard metrics in Android testing and represent a reasonable combination of fine and coarse-grained metrics.

4.6   Results

The coverage distributions for the four algorithms are illustrated in Fig. 2, whereas the left plot (Fig. 2a) refers to the activity coverage and the right plot (Fig. 2b) to the branch coverage, respectively. Concerning the medians, MIOEDA achieves the highest values with 76.41% (activity coverage) and 33.33% (branch coverage), closely followed by *Random* with 76.28% and 33.19% as well as MIO with 75.59% and 32.86%, while STOAT follows with some larger gap at 73.03% and 27.98%, respectively. Table 2 summarises the pairwise comparisons where there are statistically significant differences. To determine which configuration is significantly better, the Vargha-Delaney $\hat{A}_{12}$ effect size was used. For instance, MIOEDA outperforms STOAT with respect to activity coverage in 24 apps and 46 regarding branch coverage. Conversely, STOAT is significantly better than MIOEDA in 31 cases concerning activity coverage and in 27 cases with respect to branch coverage. Generally, MIOEDA is better than any other algorithm concerning branch coverage.

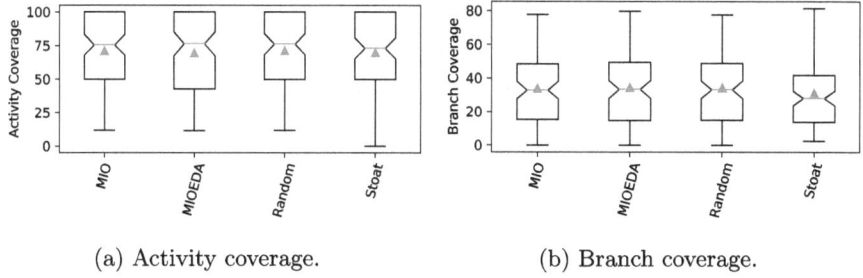

(a) Activity coverage.   (b) Branch coverage.

Fig. 2. Coverage between the different algorithms.

Table 2. Pairwise comparison of statistical significances with respect to activity and branch coverage. E.g., MIOEDA achieved significantly higher branch coverage in 46 apps than STOAT, and in 27 cases vice versa.

	Activity Coverage				Branch Coverage			
	MIOEDA	MIO	*Random*	STOAT	MIOEDA	MIO	*Random*	STOAT
MIOEDA	-	5	1	24	-	26	15	46
MIO	11	-	1	26	9	-	2	46
Random	10	2	-	27	11	13	-	45
STOAT	31	26	26	-	27	29	29	-

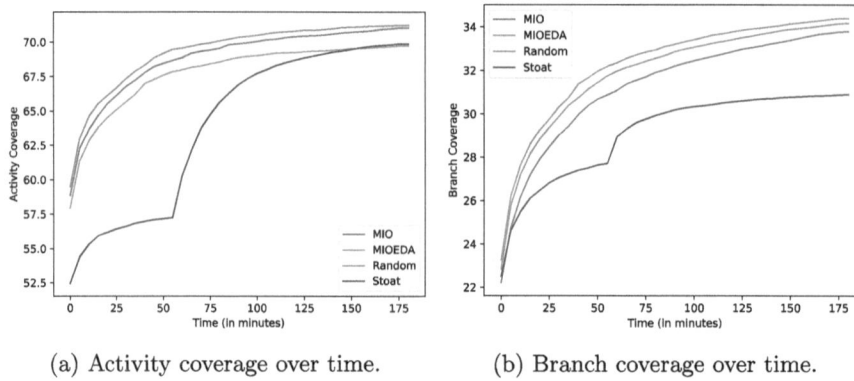

(a) Activity coverage over time.   (b) Branch coverage over time.

Fig. 3. Coverage over time.

Although it is common to see only small coverage differences between different algorithms in the Android testing domain [4,5,18], it is interesting but not surprising to note that *Random* performs almost on par with MIOEDA. With respect to activity coverage, *Random* even exhibits the highest mean over time among all algorithms (cf. Fig. 3a). This can be partially attributed to the fact that MIOEDA focuses towards covering branches, in particular branches that might reside within already covered activities, while *Random* by nature follows a completely undirected exploration. Figure 3b indeed shows that MIOEDA is ahead of *Random* concerning branch coverage over time, thus the choice of the algorithm depends on what coverage objective is more favourable in the given context. A different explanation why *Random* is almost on par with MIOEDA might be the costly fitness evaluations [4], which hamper the performance of MIOEDA. In fact, MIOEDA has to evaluate the fitness after every single action with respect to each branch and this overhead can be observed when inspecting the total number of produced tests: On average, *Random* executed 22 additional tests in comparison to MIOEDA (144 vs. 122), which corresponds to an increase of more than 18%. This rather low number of tests produced within a time frame of 3 h in comparison to other domains might further inhibit MIOEDA's full power. However, additional experiments with a search budget of 6 h showed the same trend, thus indicating this is likely of less influence. One might also conjecture that *Random* can perform extremely well on simple apps and that MIOEDA should outperform *Random* on more complex apps. We grouped the 100 apps based on its complexity, i.e., both the number of activities and branches, and plotted the coverage for each group and algorithm. However, we could not observe any clear trends that would confirm this assumption.

Comparing MIOEDA with MIO shows that MIO is slightly ahead regarding activity coverage over time, while MIOEDA performs better with respect to branch coverage. The cut point mutation employed by MIO might be responsible for the activity coverage variance, since actions are chosen randomly from the cut point onwards, thus strengthening exploration. Conversely, the higher branch coverage for MIOEDA can be attributed to the usage of a probabilistic model instead of variation operators by MIOEDA, since by construction the core algorithm is the same otherwise. An explanation why MIOEDA did not exhibit an even stronger performance in contrast to MIO could again be the lower number of produced tests (122 vs. 136). Although both algorithms need to evaluate the fitness for each objective, our implementation of MIOEDA does this repeatedly on the same execution trace when trying to distinguish good and bad nodes (cf. Subsect. 3.3). Overall, however, this nevertheless confirms the EDA as a valid alternative to classical variation operators.

We compare against STOAT since it implements an algorithm that is akin to EDAs. Here, MIOEDA is substantially better regarding both coverage criteria. The reasons can be twofold: First, MIOEDA constructs and leverages the probabilistic model from the early beginning on, while STOAT starts sampling from the probabilistic model at a later stage. This can be well observed from the coverage over time plots (cf. Fig. 3); the saddle points around 60 min refer to the moment where STOAT switches over from the model construction phase to the stochastic model sampling phase. From this point onwards, STOAT gets closer, indicating that sampling from a probabilistic model can be beneficial. The second potential reason can be the fact that MIOEDA is a many-objective algorithm, while STOAT optimises towards a linear combination of coverage and test diversity. Previous studies [3,14,15] emphasise that MOAs are superior to approaches combining objectives into a single aggregated value, and our results seem to confirm this also in the domain of Android testing.

Summary (*RQ*): Our initial investigation suggests that MIOEDA is a viable alternative for Android test generation, and performs best at optimising tests for fine-grained coverage criteria like branch coverage.

5 Related Work

The sheer number of coverage objectives and the costly test execution that would be associated with targeting objectives sequentially, led to the introduction of many-objective search algorithms like MOSA [14], DynaMOSA [15] and MIO [3]. While MOSA considers all branches of the program under test as objectives simultaneously, DynaMOSA dynamically selects the target branches based on the control dependencies among those targets. MIO was proposed based on the observation that it is difficult to cover large numbers of objectives within a limited search budget. MIOEDA approach is built on top of MIO, and mainly replaces the traditional archive with a probabilistic model originating from EDAs. Our experiments indicate that MIOEDA performs slightly better than MIO.

There are some approaches using EDAs to generate tests: Almandoz proposed in his dissertation [1] an EDA for the generation of test inputs. Similar to our approach, a control flow graph is used to select branches as possible candidates, while Wei et al. [21] leverage a combination of a Markov chain usage model and an EDA for the test data generation in the context of mutation testing. In contrast, Sagarna et al. [16] suggest an EDA for the testing of object-oriented software. In contrast, we address test generation at the system level. In this regard, STOAT [19] is related as it uses a probabilistic model in a later phase of the algorithm to sample test suites. Our approach utilises similar heuristics to assign the initial action probabilities, but in contrast to STOAT, MIOEDA constructs a probabilistic model from the early beginning on, which is iteratively refined with the information of sampled tests. Moreover, our approach is capable of optimising the search towards multiple objectives at once, while the search of STOAT is driven by a linear combination of coverage and test diversity. The results of our experiments indicate that MIOEDA outperforms STOAT, in particular in the early phase where STOAT does not yet use a probabilistic model.

6   Conclusions

Classical genetic algorithms require crossover and mutation operators to create new but diverse chromosomes. However, in the Android testing domain the creation of such variation operators can be difficult, since the individual actions that form a test are most often state-dependent and consequently cannot be arbitrarily re-arranged without potentially leading to non-executable tests. In this paper, we therefore proposed the MIOEDA search-based many-objective algorithm that requires neither a crossover nor a mutation operator. Instead, our approach builds upon an estimation of distribution algorithm (EDA) that samples new chromosomes from a probabilistic model, which is iteratively refined. In addition, to be able to handle multiple objectives within a limited search budget, we integrate the state-of-the-art many objective independent (MIO) algorithm.

Experiments with our prototype on a set of 100 F-Droid apps demonstrate that MIOEDA can compete with state-of-the-art search-based algorithms. In particular, MIOEDA achieves a median activity and branch coverage of 76.41% and 33.33%, respectively, while *Random* closely follows with 76.28% and 33.19%, which is only slightly ahead of MIO with 75.59% and 32.86%, while STOAT follows with a larger gap at 73.03% and 27.98%, respectively.

Although the results indicate that MIOEDA can be used to effectively test Android apps for coverage, there is still room for improvements. The fitness evaluations can be quite time-consuming since the fitness has to be recorded after each action in order to differentiate between good and bad actions. One could seek for a highly-parallelised evaluation. Another performance aspect is the sequential probabilistic model update procedure. In future work this could be replaced with a parallelised solution once there is enough available memory to store a dedicated model for each objective. In fact, we had to find for the prototype implementation a reasonable trade-off between memory and speed, which led among other things to the exclusion of intent actions due to the low-memory constraints on Android devices. However, previous studies [6] showed that intents can be an effective addition for testing.

To enable the replication of our results, we provide a replication package containing the study subjects, the algorithm implementation in MATE and the raw results at the following address https://doi.org/10.6084/m9.figshare.25556967.

Acknowledgment. This work is supported by DFG project FR2955/4-1 "STUNT: Improving Software Testing Using Novelty" and the COMET K1-Zentrum Competence Center for Integrated Software and AI Systems (INTEGRATE).

References

1. Almandoz, R.S.: An optimization approach for software test data generation: applications of estimation of distribution algorithms and scatter search (2007)
2. Amalfitano, D., Amatucci, N., Fasolino, A.R., Tramontana, P.: AGRippin: a novel search based testing technique for android applications. In: Proceedings of the DeMobile, pp. 5–12. ACM (2015)
3. Arcuri, A.: Many independent objective (MIO) algorithm for test suite generation. In: Menzies, T., Petke, J. (eds.) SSBSE 2017. LNCS, vol. 10452, pp. 3–17. Springer, Cham (2017). https://doi.org/10.1007/978-3-319-66299-2_1
4. Auer, M., Adler, F., Fraser, G.: Improving search-based android test generation using surrogate models. In: Papadakis, M., Vergilio, S.R. (eds.) Search-Based Software Engineering, SSBSE 2022. LNCS, vol. 13711, pp. 51–66. Springer, Cham (2022). https://doi.org/10.1007/978-3-031-21251-2_4
5. Auer, M., Pusl, M., Fraser, G.: Generating android tests using novelty search. In: Arcaini, P., Yue, T., Fredericks, E.M. (eds.) Search-Based Software Engineering, SSBSE 2023. LNCS, vol. 14415, pp. 3–18. Springer, Cham (2024). https://doi.org/10.1007/978-3-031-48796-5_1
6. Auer, M., Stahlbauer, A., Fraser, G.: Android fuzzing: balancing user-inputs and intents. In: 2023 IEEE Conference on Software Testing, Verification and Validation (ICST), pp. 37–48 (2023)
7. Eler, M.M., Rojas, J.M., Ge, Y., Fraser, G.: Automated accessibility testing of mobile apps. In: ICST, pp. 116–126. IEEE (2018)
8. Gross, F., Fraser, G., Zeller, A.: EXSYST: search-based GUI testing. In: Proceedings of the International Conference on Software Engineering, June 2012
9. Hauschild, M., Pelikan, M.: An introduction and survey of estimation of distribution algorithms. Swarm Evol. Comput. **1**(3), 111–128 (2011)
10. Kong, P., Li, L., Gao, J., Liu, K., Bissyand, T.F., Klein, J.: Automated testing of android apps: a systematic literature review. IEEE Trans. Reliab. **68**(1), 45–66 (2019)
11. Kumar, T., Iba, H.: Linear and combinatorial optimizations by estimation of distribution algorithms. In: MPS Symposium on Evolutionary Computation, January 2002
12. Mahmood, R., Mirzaei, N., Malek, S.: EvoDroid: segmented evolutionary testing of Android apps. In: Proceedings of the FSE, pp. 599–609. ACM (2014)
13. Mao, K., Harman, M., Jia, Y.: Sapienz: multi-objective automated testing for Android applications. In: Proceedings of the ISSTA, pp. 94–105. ACM (2016)
14. Panichella, A., Kifetew, F.M., Tonella, P.: Reformulating branch coverage as a many-objective optimization problem. In: 2015 IEEE 8th International Conference on Software Testing, Verification and Validation (ICST), pp. 1–10 (2015)

15. Panichella, A., Kifetew, F.M., Tonella, P.: Automated test case generation as a many-objective optimisation problem with dynamic selection of the targets. IEEE Trans. Softw. Eng. **44**(2), 122–158 (2018)
16. Sagarna, R., Arcuri, A., Yao, X.: Estimation of distribution algorithms for testing object oriented software. In: 2007 IEEE Congress on Evolutionary Computation, pp. 438–444 (2007)
17. Salustowicz, R., Schmidhuber, J.: Probabilistic incremental program evolution. Evol. Comput. **5**(2), 123–141 (1997)
18. Sell, L., Auer, M., Frädrich, C., Gruber, M., Werli, P., Fraser, G.: An empirical evaluation of search algorithms for app testing. In: Gaston, C., Kosmatov, N., Le Gall, P. (eds.) ICTSS 2019. LNCS, vol. 11812, pp. 123–139. Springer, Cham (2019). https://doi.org/10.1007/978-3-030-31280-0_8
19. Su, T., et al.: Guided, stochastic model-based GUI testing of android apps. In: Proceedings of the 2017 11th Joint Meeting on Foundations of Software Engineering, ESEC/FSE 2017, pp. 245–256. ACM (2017)
20. Vargha, A., Delaney, H.D.: A critique and improvement of the CL common language effect size statistics of McGraw and Wong. J. Educ. Behav. Stat. **25**(2), 101–132 (2000)
21. Wei, C., Yao, X., Gong, D., Liu, H.: Test data generation for mutation testing based on Markov chain usage model and estimation of distribution algorithm. IEEE Trans. Softw. Eng. **50**, 551–573 (2024)

Iterative Refactoring of Real-World Open-Source Programs with Large Language Models

Jinsu Choi⊙, Gabin An⊙, and Shin Yoo(✉)⊙

KAIST, Daejeon, Republic of Korea
{jinsuchoi,agb94,shin.yoo}@kaist.ac.kr

Abstract. Code refactoring is a critical task for improving software quality, but it is traditionally a manual, time-consuming process. This paper demonstrates an approach to automate project-level code refactoring using Large Language Models (LLMs). The key idea is to iteratively identify methods with high cyclomatic complexity, and then use LLMs to generate refactored implementations that reduce complexity. Our evaluation using 17 open-source projects shows that the proposed automated refactoring can reduce average cyclomatic complexity by up to 10.4% within 20 iterations. These results suggest that automated project-level code refactoring is feasible using LLMs with tailored prompts.

Keywords: Code Refactoring · Large Language Model · Cyclomatic Complexity

1 Introduction

Code refactoring is the process of modifying a software system's internal structure without changing its external behavior [1]. Traditionally, code refactoring tasks are performed manually by developers based on their experience and industry best practices [2], which can make large-scale refactoring a time-consuming process that takes weeks to months to complete.

Recently, there has been growing interest in applying Large Language Models (LLMs) to various Software Engineering tasks, such as code generation, test generation, and automated debugging [3]. Prior work [4,5] has noted that LLMs can be effectively combined with traditional search-based software engineering techniques because this combination allows LLMs to provide more powerful and tailored code mutations, while the generate-and-validate approach helps prevent LLMs from generating unreliable or hallucinated outputs. Code refactoring is particularly well-suited for leveraging LLMs, as the assurance criteria can be fully automated using a regression test oracle [4]. However, existing LLM-based refactoring work has relied only on existing tests to check the correctness of refactoring patches, without the use of such regression oracles [6]. Additionally, this study has evaluated performance using only introductory programs, rather than real-world software projects.

G. Jahangirova and F. Khomh (Eds.): SSBSE 2024, LNCS 14767, pp. 49–55, 2024.
https://doi.org/10.1007/978-3-031-64573-0_4

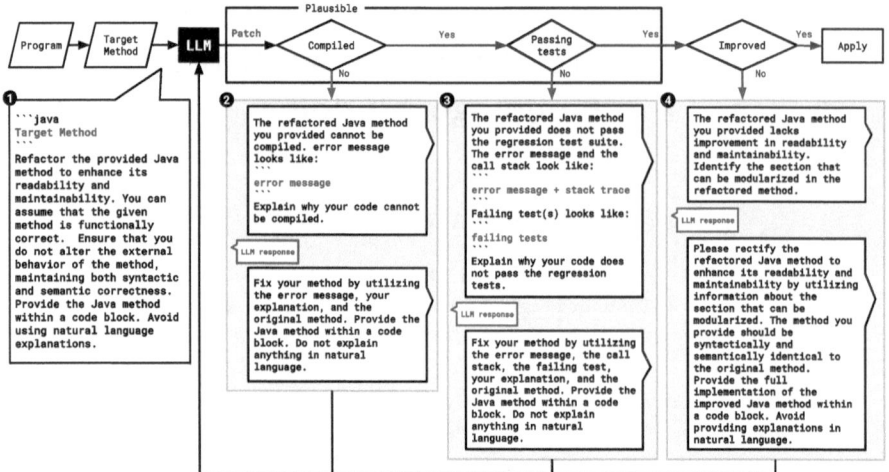

Fig. 1. Overview

This paper demonstrates the use of LLMs to perform iterative project-level code refactoring with an objective of reducing the Cyclomatic Complexity (CC) [7], the number of linearly independent paths within the source code. The approach first identifies the methods with the highest CC and then performs refactoring on them. Once an improved version of the method implementation is found, it is applied to the codebase, and the approach subsequently attempts to refactor the next high-complexity method. To prevent the existing functionality from breaking during the refactoring, we use automated generated regression tests to filter out incorrect patches and best assure that the refactored code maintains the original behavior [4]. To evaluate our approach, we use 17 open-source projects from the Defects4J benchmark [8]. The results demonstrate that our method can reduce the average CC of the programs by up to 10.4% within 20 iterations. We observe that LLMs sometimes significantly reduce the CC by splitting a complex method into multiple simple methods. Overall, these findings suggest the viability of LLM-based automated refactoring for large-scale software projects.

2   Methodology

Our methodology for iteratively reducing the code complexity of a target program is illustrated in Fig. 1. The process begins by identifying the method with the highest CC in the project. Subsequently, the LLM is requested to refactor this selected method (Sect. 2.1). The refactored code passes through a series of filters. To ensure that the refactoring does not break the existing functionality of the original code, the plausibility of the refactored code is verified using two types of regression tests: developer-written tests and automatically generated tests (Sect. 2.2). If the LLM has indeed improved the code quality, the refactored code is applied to the project (Sect. 2.3). This iterative process is repeated, progressively enhancing the overall code quality of the project.

2.1 Extract and Refactoring Target Method

We use the Lizard library[1] to measure the CC of all methods within the source files of the project. After identifying the method with the highest CC, we retrieve the source code of that particular method. The retrieved method implementation is then embedded into the prompt (❶ in Fig. 1) which is provided as input to the LLM. The prompt instructs the LLM to improve the readability and maintainability of the method implementation, while ensuring that the refactored method remains semantically equivalent to the original one.

2.2 Checking Plausibility

After the LLM generates the refactoring patch, we verify the plausibility of the refactored method implementation. This involves checking whether the program is still compilable and whether it successfully passes both the developer-written test suite and the automatically generated regression tests. Note that the regression tests were created with respect to the original program before any refactoring was applied. If the refactored method is implausible, we request the LLM to rectify the issues. In case of a compilation failure, the prompt supplies the compilation error message to the LLM (❷ in Fig. 1). If the compilation succeeded but the tests did not pass, the prompt provides both a stack trace and details of the failing tests, in addition to the error message (❸ in Fig. 1). After presenting this error information, the prompt requires the LLM to analyze the root cause of the issue. Subsequently, the prompt requests the LLM to fix the method, i.e., regenerate the refactoring patch. If the fixed method remains implausible, meaning it still fails the checks, we exclude that method from the improvement targets and proceed to the next iteration without applying the refactored method in the project.

2.3 Assessing Improvement

If a plausible patch is found, we measure the CC of the refactored method and compare it to the CC of the original method. If the LLM has split the original method into multiple methods, we compare the CC of the method with the highest complexity. If the CC is reduced, the refactored method is then applied to the target project, and we move forward with the next iteration. This iterative process allows the project to be continuously improved. If the CC of the refactored method does not decrease, i.e., there is no improvement in complexity, we request the LLM to identify a part of the code that can be further modularized (❹ in Fig. 1). Subsequently, the LLM is tasked with enhancing the quality of the code based on its own analysis and response. If the modified method is

---

[1] https://github.com/terryyin/lizard.

Table 1. Refactoring results: the percentage reduction in the average CC, the percentage increase in the number of functions, and the percentage reduction in the average number of lines of code. The maximum values across the five attempts are presented.

Project	Avg. CC Reduction	# Func Increase	Avg. nLoC Reduction	Project	Avg. CC Reduction	# Func Increase	Avg. nLoC Reduction
Chart	0.27%	0.10%	0.74%	Cli	4.94%	5.70%	4.67%
Closure	0.35%	0.24%	0.21%	Codec	2.37%	4.41%	2.94%
Collections	0.06%	0.08%	0.06%	Compress	1.15%	0.32%	0.96%
Csv	10.40%	17.86%	10.60%	Gson	2.95%	3.32%	2.46%
JacksonCore	0.38%	0.54%	0.41%	JacksonDatabind	0.10%	0.11%	0.11%
JacksonXml	2.31%	3.10%	2.79%	Jsoup	0.88%	1.44%	1.03%
JxPath	0.79%	0.58%	0.99%	Lang	0.80%	0.67%	1.22%
Math	0.01%	0.02%	0.01%	Mockito	0.78%	0.97%	0.53%
Time	0.84%	0.42%	0.35%				

still not plausible or fails to improve the complexity, we exclude that method from future iterations and proceed to the next iteration without applying the refactored method in the project.

3 Experimental Setup

We evaluate our LLM-based refactoring pipeline using the 17 real-world Java projects from Defects4J v2.0.0 [8]. As multiple snapshots of each project are available in the benchmark, we utilize the latest version for every project. For our experiment, we used the `gpt-3.5-turbo-0125`. Due to the stochastic nature of the LLM querying process, we run our pipeline five times for each project, with each execution comprising 20 iterations of refactoring. To generate regression tests for the method in focus at each iteration, we make use of the *gen_tests* script of Defects4J. In particular, EvoSuite [9] is utilized to generate tests for Java classes that contain the target method, with an allocated time budget of 180 s. If the generated regression tests lead to failures when run against the program, we eliminate those test cases and initiate the test generation process again, allowing for a maximum of five attempts. If failures persist beyond five attempts, it is assumed that the target method (or class) contains elements of non-determinism, leading us to halt further attempts to improve the method.

4 Results

Refactoring Results: We examine the extent to which the average CC of the entire project decreases as the iteration progresses. Figure 2 illustrates the change in average CC across projects over successive iterations. In every project, there is at least one instance where the average CC decreases. On average, our method reduces the average CC of the project by 1.2%. Table 1 displays more

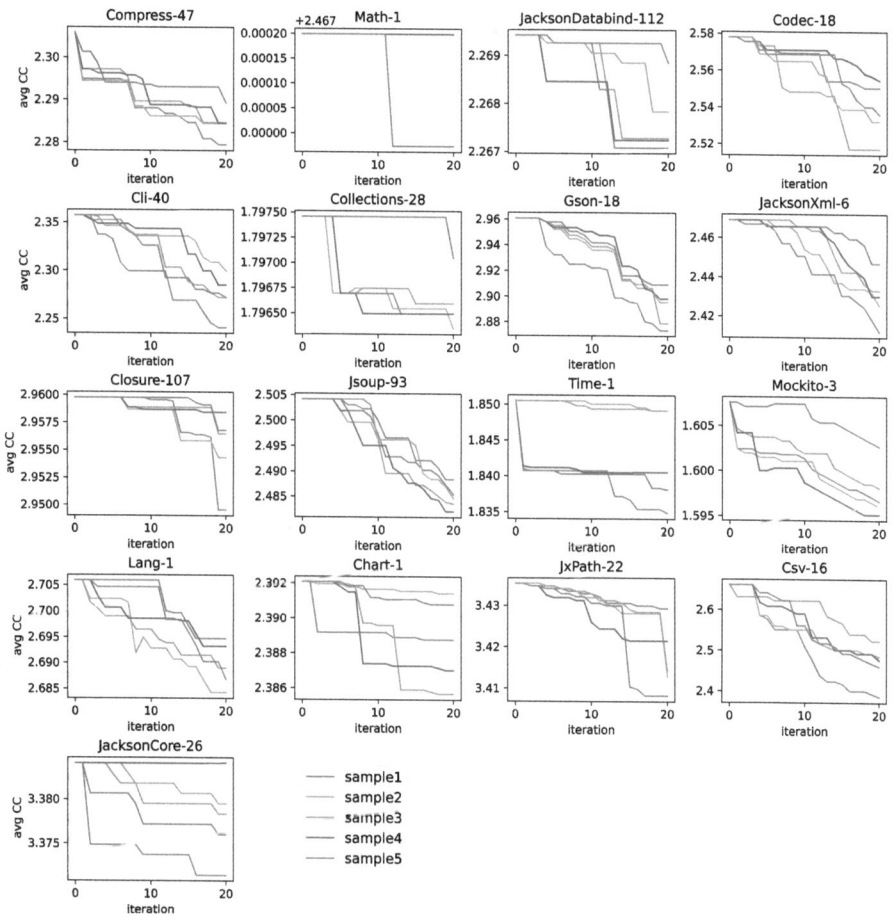

Fig. 2. Change in average CC over 20 iterations

detailed refactoring results after the 20 iterations. The highest reduction rate of average CC is 10.4% for Csv-16. During the refactoring process, there are instances where the target method is split into multiple methods, resulting in the number of functions in the project increasing by 1.5% on average across all refactoring attempts. Notably, in Csv-16, where the average CC decreased the most, the number of functions increased by 17.9%. As the method is separated, the average length of the method is shortened: the average number of Lines of Code without comments (nLoC) is reduced by 1.3% on average. The examples of LLM-generated refactoring patches are available online at https://figshare. com/s/6e7d9f69a96974c110fb.

Detailed Statistics of the LLM-Generated Code: During our experiment, the LLM is requested to refactor the method a total of 1,700 times across all projects. We found that 45.5% of the refactoring attempts initially generate non-plausible methods. Among them, 14.0% of the non-plausibility can only be discovered through automated regression tests. The plausibility fixing process leads to 33.7% of the initially non-plausible methods becoming plausible. After this fixing process, 69.8% of all attempts produce plausible methods. However, among these, 81.7% show no improvement. Further modularization process leads to improvement in 30.9% of these initially non-improved methods.

5   Conclusion and Future Work

In this study, we explore the potential of using LLMs to enhance the quality of program source code, specifically aiming to lower CC. We discovered that integrating LLMs with our iterative search methodology and tailored prompts successfully decreases the project's average CC and enhances its modularity, all while preserving its original functionality, which is confirmed through both developer-written and automated regression testing. However, manual verification is still necessary to ensure that the code remains semantically identical. Furthermore, our current pipeline has been limited to the sequential refactoring of individual methods. Moving forward, we aim to develop a more structural refactoring approach using LLMs, such as the detection and modularization of recurring code patterns across different methods as well as operations that involve multiple methods. We also plan to assess the effectiveness of LLM-based refactoring on other non-functional quality metrics, such as execution time and memory usage. Finally, we will investigate whether LLMs can guide more qualitative refactoring goals, instead of simply following quantitative metrics [10].

References

1. Fowler, M.: Refactoring: Improving the Design of Existing Code. Addison-Wesley, Boston (1999)
2. Murphy-Hill, E., Parnin, C., Black, A.P.: How we refactor, and how we know it. IEEE Trans. Softw. Eng. **38**(1), 5–18 (2012)
3. Fan, A., et al.: Large language models for software engineering: survey and open problems. arXiv preprint arXiv:2310.03533 (2023)
4. Alshahwan, N., Harman, M., Harper, I., Marginean, A., Sengupta, S., Wang, E.: Assured LLM-based software engineering (2024)
5. Kang, S., Yoo, S.: Towards objective-tailored genetic improvement through large language models. In: 2023 IEEE/ACM International Workshop on Genetic Improvement (GI), pp. 19–20, IEEE (2023)
6. Shirafuji, A., Oda, Y., Suzuki, J., Morishita, M., Watanobe, Y.: Refactoring programs using large language models with few-shot examples. arXiv preprint arXiv:2311.11690 (2023)
7. McCabe, T.: A complexity measure. IEEE Trans. Softw. Eng. **SE–2**(4), 308–320 (1976)
8. Just, R., Jalali, D., Ernst, M.D.: Defects4J: a database of existing faults to enable controlled testing studies for java programs. In: Proceedings of the 2014 International Symposium on Software Testing and Analysis, ISSTA 2014, New York, NY, USA, pp. 437-440, Association for Computing Machinery (2014)

9. Fraser, G., Arcuri, A.: EvoSuite: automatic test suite generation for object-oriented software. In: Proceedings of the 19th ACM SIGSOFT Symposium and the 13th European Conference on Foundations of Software Engineering, ESEC/FSE 2011, New York, NY, USA, pp. 416–419. ACM (2011)
10. Simons, C., Singer, J., White, D.R.: Search-based refactoring: metrics are not enough. In: Barros, M., Labiche, Y. (eds.) SSBSE 2015. LNCS, vol. 9275, pp. 47–61. Springer, Cham (2015). https://doi.org/10.1007/978-3-319-22183-0_4

Approximating Stochastic Quantum Noise Through Genetic Programming

Asmar Muqeet[1,2]([⊠]) [ID], Shaukat Ali[1,3]([⊠]) [ID], and Paolo Arcaini[4] [ID]

[1] Simula Research Laboratory, Oslo, Norway
{asmar,shaukat}@simula.no
[2] University of Oslo, Oslo, Norway
[3] Oslo Metropolitan University, Oslo, Norway
arcaini@nii.ac.jp
[4] National Institute of Informatics, Tokyo, Japan

Abstract. Quantum computing's potential for exponential speedups over classical computing has recently sparked considerable interest. However, quantum noise presents a significant obstacle to realizing this potential, compromising computational reliability. Accurate estimation and mitigation of noise are crucial for achieving fault-tolerant quantum computation. While current efforts focus on developing noise models tailored to specific quantum computers, these models often fail to fully capture the complexity of real quantum noise. To this end, we propose an approach that uses genetic programming (GP) to develop expression-based noise models for quantum computers. We represent the quantum noise model as a computational expression, with each function corresponding to a specific aspect of the noise behavior. By function nesting, we create a chain of operations that collectively capture the intricate nature of quantum noise. Through GP, we explore the search space of possible noise model expressions, gradually improving the quality of the solution. We evaluated the approach on five artificial noise models of varying complexity and a real quantum computer. Results show that our approach achieved an error difference of less than 2% in approximating artificial noise models and 15% for a real quantum computer.

Keywords: quantum noise · quantum computing · genetic programming

1 Introduction

In recent years, Quantum Computing (QC) has gained significant attention due to its potential speed advantage over classical computing in solving specific problem classes more efficiently [1]. However, one of the major hurdles to achieving

The work is supported by the Qu-Test project (Project #299827) funded by the Research Council of Norway. S. Ali is also supported by Oslo Metropolitan University's Quantum Hub. P. Arcaini is supported by Engineerable AI Techniques for Practical Applications of High-Quality Machine Learning-based Systems Project (Grant No JPMJMI20B8), JST-Mirai.

G. Jahangirova and F. Khomh (Eds.): SSBSE 2024, LNCS 14767, pp. 56–62, 2024.
https://doi.org/10.1007/978-3-031-64573-0_5

quantum advantage is *quantum noise*, which adversely affects the computation of quantum computers, leading to undesired behaviors. Estimating noise in quantum computers has become crucial, as we are now testing novel quantum error correction and quantum error mitigation methods on current *noisy intermediate-scale quantum* (NISQ) devices [8]. Current efforts in noise estimation focus on identifying *noise models* that accurately represent the noise errors present in current NISQ computers [5,6]. However, existing noise models are simple approximations of real quantum noise [6]. A more detailed noise model that not only represents the noise errors in a quantum computer but also depicts the relationship between different noise errors, qubits, and gate operations would be immensely beneficial. Such a model would greatly enhance the precision of quantum error correction and mitigation techniques. Studies have shown that knowing a more detailed noise model for each qubit and gate operation for a particular quantum computer can significantly enhance the accuracy of quantum error mitigation methods [11,13].

In NISQ computers, noise has various forms, e.g., depolarizing, amplitude dampening, and phase dampening noise [4]. Each qubit and gate operation may experience multiple noise errors, which vary for different qubits and gate operations. In this paper, we propose an approach that uses *genetic programming* (GP) to create expression-based noise models for specific NISQ computers. We represent the quantum noise model as a computational expression consisting of a chain of function calls. Each function adds a specific noise error to particular qubits and gate operations. By nesting these functions within a computational expression, we capture the intricate relationships among various noise errors, qubits, and gate operations for specific quantum computer configurations. The GP process begins with an initial random population of candidate noise model expressions. Through the proposed fitness function and standard evolutionary operators, new candidates are evaluated and generated. This iterative process explores the search space of potential noise model expressions, gradually improving the quality of solutions until satisfactory noise models are obtained. Importantly, by representing noise models as computational expressions, their complex mathematical representation is abstracted, enhancing human comprehension. We evaluate our approach by approximating five artificially created noise models with varying strengths and approximating the noise of one real NISQ computer, IBM-Kyoto. Our approach approximated artificial noise models with less than a 2% difference across all models. Moreover, for IBM-Kyoto, our method outperformed the baseline, with a 15% difference in the noise model approximation compared to 40% for the baseline. In summary, our contributions are (1) the application of GP for approximating quantum noise and creating a more interpretable noise model; (2) an empirical evaluation with five artificial noise models and evaluating the applicability on a real quantum computer.

Related Work. Several efforts have been made to create noise models for NISQ computers. For example, Harper et al. [5] proposed a noise estimation method for quantifying noise in quantum systems and creating correlation matrices that describe the relationship of errors with different qubits. Harper et al. [6] proposed an algorithm for creating a noise model of sparse Pauli noise errors for Clifford quantum circuits. Moreover, Georgopoulos et al. [4] proposed the use of noise estimation circuits to model depolarizing noise error for a given quantum circuit, and Wise et al. [12] used deep learning to transform the output of a quantum circuit to a noisy output, resulting in a neural network-based noise model. However, the major limitation of all these methods is that they are difficult to analyze due to the closed nature of machine learning models and quantum states and, thus, cannot be directly utilized by quantum error mitigation methods.

2 Approach

The most common errors due to quantum noise are depolarizing, amplitude-dampening, and phase-dampening errors [4]. **Depolarizing Error** arises from the interaction of a quantum computer with its environment and describes the probabilistic process by which the quantum state of a qubit undergoes random rotations or flips, leading to computation errors [7]. Formally, for one qubit, it is defined as

$$\mathcal{D}(\rho) = (1 - p)\rho + \frac{p}{3} \left(X\rho X + Y\rho Y + Z\rho Z \right) \tag{1}$$

where p is the probability of the error, ρ is the density matrix, and X, Y, Z are the Pauli operations. Different qubits and gate operations can have different probabilities of the depolarizing error [7]. **Amplitude damping** error refers to the loss of energy from a quantum system to its environment, while **Phase damping** error refers to the loss of information from a quantum system to its environment without dissipating energy. Formally, both are represented as

$$\mathcal{E}(\rho) = E_0 \rho E_0^\dagger + E_1 \rho E_1^\dagger \tag{2}$$

where ρ is the density matrix, E_0 and E_1 are the Kraus operators. For amplitude damping, E_0 is represented by the matrix $\begin{bmatrix} 1 & 0 \\ 0 & \sqrt{1-\gamma} \end{bmatrix}$, and E_1 by $\begin{bmatrix} 0 & \sqrt{\gamma} \\ 0 & 0 \end{bmatrix}$. For phase damping, E_1 is $\begin{bmatrix} 0 & 0 \\ 0 & \sqrt{\gamma} \end{bmatrix}$ and E_0 is the same as amplitude damping. These matrices depict the potential outcomes of the damping process. γ is the damping parameter, which represents the probability of the qubit transitioning from the excited state to the ground state. Our genetic programming (GP)-based approach uses these three noise errors to create an individual representing a noise model.

Individual Representation. GP uses an evolutionary algorithm to evolve computer programs, that are represented and stored as syntax trees. These trees consist of interior nodes representing operations and terminal nodes representing inputs or parameters for these operations. In GP, operation nodes are denoted by a tuple (*func*, *arity*), where *func* defines the operation and *arity* specifies the number of arguments it can take. The arity of operation nodes in the syntax tree determines the number of child nodes each operation node can have. Our approach employs a variant called Strongly Typed Genetic Programming (STGP) [9]. In STGP, operation nodes additionally define the data type of their arguments and the return type of the operation, represented as (*func*, *arity*, *argType*$_1$, ..., *argType*$_n$, *retType*).

To construct a noise model for a quantum computer, we define operation nodes in STGP corresponding to common basis gates (such as rx, ry, rz, sx, cx) supported by real quantum computers. Figure 1 illustrates an example individual for STGP. In this

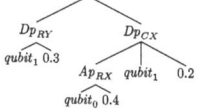

Fig. 1. An example individual for GP

figure, *Init* is the root node representing the initialization of an empty noise model. Dp_{RY} is an operation node taking two arguments: the qubit index and the probability p for a depolarizing noise error (see Eq. 1). For instance, Dp_{RY} with arguments $qubit_1$ and 0.3 indicates a depolarizing noise error on qubit 1 and gate operation ry with a probability of 0.3. The return value of each operation node, such as $(Dp_{RX}, Dp_{CX}, Ap_{RX})$, is the qubit number it acted upon. This enables combining different noise errors on a particular qubit and gate operation to create a more accurate representation of actual noise. For two-qubit gate operations like Dp_{CX}, it takes two qubit indexes as arguments, where the first index is the control qubit and the second index is the target qubit. The return value of two-qubit operation nodes is the index of the target qubit. One advantage of GP is its ability to convert the entire tree representation into a computational expression, defining how noise affects different qubits and gate operations. For instance, the individual in Fig. 1 can be converted into the following computational expression:

$$(Init \; (Dp_{RY} \; qubit_1 \; 0.3) \; (Dp_{CX} \; (Ap_{RX} \; qubit_0 \; 0.4) \; qubit_1 \; 0.2))$$

By utilizing Eqs. 1 and 2, this expression can be translated into the following noise representation from a qubit perspective:

$$Q_0 = \mathcal{E}(\rho)_{0.3} \qquad\qquad Q_1 = \mathcal{D}(\rho)_{0.3} \otimes (\mathcal{D}(\rho)_{0.2}|Q_0)$$

Fitness Function. Fitness is calculated by averaging the Hellinger distance between multiple quantum circuits. The Hellinger distance is widely used for assessing the output of quantum circuits under noise [2]. The fitness function is $\frac{1}{n}\sum_{i=1}^{n}\frac{1}{\sqrt{2}}|\sqrt{P_i} - \sqrt{X_i}|$, where n is the number of circuits used, P_i is the output of the i-th circuit from the real computer, and X_i is the output of the i-th circuit under the noise model. We utilize multiple quantum circuits to evaluate the fitness to avoid optimizing a noise model for a specific quantum circuit.

3   Experiment Design and Result

For our experiment[1], we implemented STGP using the DEAP framework [3], with default settings. By default, DEAP uses a half-and-half policy for initialization, one-point crossover and uniform mutation for genetic variations, and the tournament selection method for choosing the best individuals[2]. For the initialization policy, we set the minimum depth to zero and the maximum depth

---

[1] https://doi.org/10.5281/zenodo.11198788.
[2] https://deap.readthedocs.io/en/master/api/tools.html.

Table 1. Comparison of 10 runs of genetic programming with baseline

Noise Model	STGP		Random		Statistics	
	F_{avg}	F_{std}	F_{avg}	F_{std}	p_{value}	$\hat{A}_{12}$
Af1	**0.0024**	0.0014	0.44	0.0009	0.0002	Large
Af2	**0.0196**	0.0156	0.38	0.0013	0.0002	Large
Af3	**0.0054**	0.0021	0.21	0.0013	0.0002	Large
Af4	**0.0018**	0.0019	0.29	0.0011	0.0002	Large
Af5	**0.0007**	0.0006	0.20	0.0008	0.0002	Large
IBM-Kyoto	**0.148**	6.6e−05	0.41	0.0006	0.0002	Large

to two, and for the selection method, we set the tournament size to three. For quantum program execution and noise model creation, we utilized IBM's Qiskit framework.

As benchmarks, we created five random noise models with varying complexity, each composed of combinations of three selected noise errors. We evaluated our approach on a real quantum computer from IBM (IBM-Kyoto). Fitness was calculated using three quantum circuits (Amplitude Estimation, Phase Estimation, and Quantum Fourier Transform), computing the average Hellinger distance value, which ranges from 0 to 1, indicating no difference to maximum difference, respectively. To assess effectiveness, we compared our approach with a random baseline across 10 repeated runs. For statistical analysis, we use the Mann-Whitney test and Vargha Delaney $\hat{A}_{12}$ effect size as recommended in [10]. $\hat{A}_{12}$ is interpreted according to [10]: an effect size in the range $(0.34, 0.44]$ and $[0.56, 0.64)$ is considered *Small*; in $(0.29, 0.34]$ and $[0.64, 0.71)$ is considered *Medium*; in $[0, 0.29]$ and $[0.71, 1]$ is considered *Large*. STGP ran for 40 generations with a population size of 300, using default parameter values from DEAP for all other parameters. For a baseline comparison, we generated 12k random individuals, aligning with the generation and population size of STGP.

Result. Table 1 presents the results, where the columns F_{avg} and F_{std} indicate the average, and standard deviation of the best individual over 10 repeated runs. The best average values, highlighted in bold, signify a closer approximation to zero, indicating a better fit of the noise model. Our approach outperformed the baseline random method for all five artificial noise models, with statistically significant improvements indicated by a p_{value} of less than 0.05 and a $\hat{A}_{12}$ statistics with *Large* magnitude. It achieved average fitness values of less than 2%, with consistently low standard deviation across all models. For the real quantum computer (IBM-Kyoto), our approach outperformed the baseline with an average fitness of 15% compared to the 40% for baseline. The results demonstrate our approach's effectiveness in approximating the noise model based on program output for both artificial noise model and real quantum computer. This demonstrates that our approach effectively approximates the noise model of a real quantum computer using the expression representation of GP.

Limitations. While our approach effectively approximates the noise model, there was a notable difference between the results of artificial noise models and the real quantum computer. This is because our method considers only three types of noise errors, while real quantum computers may have additional error types like readout and random unitary errors. To enhance accuracy, we plan to include these additional error types. Additionally, utilizing only three quantum circuits for fitness calculation and using the same circuits for experiment evaluation limits generalization. Therefore, adding more circuits could provide a more comprehensive evaluation of the noise model.

4 Conclusion

We present an approach that uses genetic programming to generate expression-based noise models tailored for NISQ computers. By representing noise as a chain of function calls, our approach creates interpretable noise models that capture different noise errors affecting individual qubits and gate operations of a quantum computer. Our results demonstrate the effectiveness of our approach, achieving an approximation error below 2% for artificially generated noise models and 15% for real quantum computer noise. In the future, we aim to enhance our approach by incorporating additional noise errors and expanding the range of evaluated quantum circuits to gauge its effectiveness further.

References

1. Classical and quantum computing. In: Quantum Information. Springer, New York (2007). https://doi.org/10.1007/978-0-387-36944-0_13
2. Dasgupta, S., Humble, T.S.: Characterizing the reproducibility of noisy quantum circuits. Entropy **24**(2), 244 (2022)
3. Fortin, F.A., De Rainville, F.M., Gardner, M.A.G., Parizeau, M., Gagné, C.: DEAP: evolutionary algorithms made easy. J. Mach. Learn. Res. **13**(1), 2171–2175 (2012)
4. Georgopoulos, K., Emary, C., Zuliani, P.: Modeling and simulating the noisy behavior of near-term quantum computers. Phys. Rev. A **104**, 062432 (2021)
5. Harper, R., Flammia, S.T., Wallman, J.J.: Efficient learning of quantum noise. Nat. Phys. **16**(12), 1184–1188 (2020)
6. Harper, R., Yu, W., Flammia, S.T.: Fast estimation of sparse quantum noise. PRX Quant. **2**, 010322 (2021)
7. King, C.: The capacity of the quantum depolarizing channel. IEEE Trans. Inf. Theor. **49**(1), 221–229 (2003)
8. Martinis, J.M.: Qubit metrology for building a fault-tolerant quantum computer. npj Quant. Inf. **1**(1), 1–3 (2015)
9. Montana, D.J.: Strongly typed genetic programming. Evol. Comput. **3**(2), 199–230 (1995)
10. Tomczak, M., Tomczak, E.: The need to report effect size estimates revisited. An overview of some recommended measures of effect size. Trends Sport Sci. **21**(1), 19–25 (2014)

11. Urbanek, M., Nachman, B., Pascuzzi, V.R., He, A., Bauer, C.W., de Jong, W.A.: Mitigating depolarizing noise on quantum computers with noise-estimation circuits. Phys. Rev. Lett. **127**, 270502 (2021)
12. Wise, D.F., Morton, J.J., Dhomkar, S.: Using deep learning to understand and mitigate the qubit noise environment. PRX Quant. **2**, 010316 (2021)
13. Wood, C.J.: Special session: noise characterization and error mitigation in near-term quantum computers. In: 2020 IEEE 38th International Conference on Computer Design (ICCD), pp. 13–16 (2020)

Fuzzing-Based Differential Testing
for Quantum Simulators

Daniel Blackwell[1](✉) , Justyna Petke[1] , Yazhuo Cao[2] ,
and Avner Bensoussan[2]

[1] University College London, London, UK
{daniel.blackwell.14,j.petke}@ucl.ac.uk
[2] King's College London, London, UK
{yazhuo.cao,avner.bensoussan}@kcl.ac.uk

Abstract. Quantum programs are hard to develop and test due to their
probabilistic nature and the restricted availability of quantum computers. Quantum simulators have thus been introduced to help software
developers. There are, however, no formal proofs that these simulators
behave in exactly the way that real quantum hardware does, which could
lead to errors in their implementation. Here we propose to use a search-
based technique, grammar-based fuzzing, to generate syntactically valid
quantum programs, and use differential testing to search for inconsistent
behaviour between selected quantum simulators. We tested our approach
on three simulators: Braket, Quantastica, and Qiskit. Overall, we generated and ran over 400k testcases, 2,327 of which found new coverage,
and 292 of which caused crashes, hangs or divergent behaviour. Our
analysis revealed 4 classes of bugs, including a bug in the OpenQASM
3 `stdgates.inc` standard gates library, affecting all the simulators. All
but one of the bugs reported to the developers have been already fixed
by them, while the remaining bug has been acknowledged as a true bug.

Keywords: Differential Testing · Fuzzing · Quantum Simulators

1 Introduction

Quantum computers offer an exciting opportunity to massively speed up existing computation. However, writing valid quantum programs is non-trivial. By
shifting the paradigm from traditional computing the outputs are no longer
deterministic. It is thus no wonder that testing quantum software is a challenging task [1]. Furthermore, quantum computers are not widely available, require
specialist knowledge to run and maintain, and suffer with inaccuracy due to
noise.

Quantum simulators have been introduced to ease developers in programming
and validating quantum circuits. Nevertheless, there are no formal guarantees
that the outputs of simulation will be the same on real quantum hardware.
Furthermore, simulators themselves might not be bug-free. In fact, Wang et al. [5]

G. Jahangirova and F. Khomh (Eds.): SSBSE 2024, LNCS 14767, pp. 63–69, 2024.
https://doi.org/10.1007/978-3-031-64573-0_6

generated semantics-preserving gate transformations and showed that in 33 of 730 cases the outputs of simulations diverged. The technique, however, requires specialist domain knowledge to create these metamorphic relations; we lack the knowledge to ascertain whether they have already found all of the viable ones.

To aid the development of quantum simulators we propose to use search-based differential testing to check their validity. In particular, we investigate the difference in behaviour between different quantum simulators, when given the same quantum programs, generated by a grammar-based fuzzer. As far as we know, **this is the first time our differential search-based testing approach has been applied to test quantum program simulators.**

Our initial results are encouraging. We found 4 classes of bugs, with at least one in all three quantum program simulation frameworks tested, i.e., Qiskit[1], Braket[2] and Quantastica[3]. **All bugs have been confirmed and all but one already fixed** by the developers. Furthermore, our method has generated a large number of testcases (over 400k) covering a lot of functionality which we have minimised to a set of 842 testcases that achieve 100% of the coverage discovered by the fuzzer; this may be useful to the simulator maintainers as a standalone regression test set, or alternatively our approach could be trivially adapted to do regression fuzz testing. Moreover, our methodology can aid in extending existing quantum program benchmarks. To allow further uptake of our approach we provide the artefact at https://doi.org/10.5281/zenodo.11002154 and the GitHub repo: https://github.com/GloC99/fuzzingQuantumSimulator.

2   Differential Fuzz Testing

Here we propose to perform differential fuzz testing of quantum simulators. We first take existing quantum programs representing valid circuits. We compile them into an intermediate representation: the OpenQASM 3 language [2] (herein referred to as QASM). We feed these programs to a search-based automated test generation tool to generate more testcases. We chose to use an existing grey-box fuzzer for this purpose. Grey-box fuzzing is a search-based software testing technique that generates new testcases by mutating existing ones; coverage feedback is used to retain any new testcases that exercise new program functionality. In essence, the mechanism of a grey-box fuzzer can be likened to a genetic algorithm whose fitness function is the total coverage of the retained set of testcases. We have extended the fuzzer with a QASM grammar-aware mutator. This way we generate only syntactically valid programs. Finally, we feed the generated QASM programs through different quantum simulators, observe crashes or hangs where they occur, and compare the outputs where they don't.

In order to evaluate our approach we built tooling based on the AFL++ fuzzer [3] and tested it on 3 quantum simulation frameworks: Qiskit, Braket and Quantastica. Next, we detail each step of the implementation of our approach.

---

[1] https://github.com/Qiskit/qiskit-aer.
[2] https://github.com/amazon-braket/amazon-braket-default-simulator-python.
[3] https://github.com/quantastica/quantum-circuit.

Grammar Mutator. We use `Grammar-Mutator`[4] from AFL++'s set of included custom mutators. As it takes grammars provided in an unusual JSON format; we had to manually adapt the ANTLR4 grammar from the OpenQASM 3 specification[5]. Additionally, we found that the specification is very much forward-looking, and 10 of the 28 statements are not yet implemented in the two simulators supporting OpenQASM 3. As such, these statements along with certain other unimplemented features were removed from our grammar in order to increase the number of valid executable programs we could generate. `Grammar-Mutator` works on a context-free grammar, which means that while it generates programs that will pass the lexing stage, many of these will not successfully parse due to invalid semantics. To increase the likelihood of successful parsing, and thus execution, we adjusted the grammar so that the generated program is guaranteed to begin with the declaration of a quantum register, and will include at least one gate statement; without these, the Braket simulation framework we use would throw an exception to indicate that the program has no functionality.

Instrumentation. We chose to use `python-afl`[6] in combination with AFL++, in order to have easy access to a grammar mutator. As noted in GitHub issue no. 25 by the author, the instrumentation method used by `python-afl` incurs significant runtime overhead. Running the three simulators with example programs took around 0.3–0.5 s already; with instrumentation this could be up to 2 or more seconds. To mitigate this, we added functionality to `python-afl` to allow us to enable and disable instrumentation at points in the fuzzing harness so as to eliminate the overhead of instrumentation on 'boring' functionality. This allowed us to average approximately 1.1 executions per second – still slow by fuzzing standards, but several times better than the naive approach.

It is worth noting that the Quantastica `quantum-circuit` simulator is written in JavaScript and hence cannot be invoked directly from the Python fuzzing harness. We built a Node JS server that is run on localhost, with a single endpoint to receive a QASM program as a string, load it in, execute it and send the result as a response. The fuzzing harness sends a web request and parses the response in order to compare with that of the other simulators; as a result, no coverage feedback is available for this simulator.

Fuzzing Harness. A single fuzzing harness was created that took the OpenQASM 3 input generated by the fuzzer, and fed it through all three simulators in the following order: Braket, Qiskit then Quantastica; collecting the state vector for each. As Quantastica only supports OpenQASM 2 (rather than the OpenQASM 3 that our generated programs are in), we chose to export the circuit that we have just generated in Qiskit to OpenQASM 2; this is not an ideal solution as any parsing errors made by Qiskit will be passed along, but it is the best option we have short of not testing Quantastica. Chaining the executions in the way we do means that if a crash occurs in Braket, then the input will not be

---

[4] https://github.com/AFLplusplus/Grammar-Mutator.
[5] https://openqasm.com/versions/3.0/grammar/index.html.
[6] https://github.com/jwilk/python-afl.

ran for Qiskit or Quantastica; when triaging the discovered crashes we run the generated programs on the other simulators to ensure that the same crash does not occur for them.

Any generated programs that parse and execute successfully on all simulators make it to the final step, which is comparison of the output probability distributions. For this we used Jensen-Shannon divergence [4], which is a finite symmetric measure of divergence between two probability distributions. Using an assertion, we set an arbitrary cap of 0.01 to be allowed when comparing Braket with Qiskit, and Qiskit with Quantastica; failing to pass this assertion marks this input as a crash in the fuzzer. We could set such a strict cap on expected divergence because directly using the state vectors avoids noise and should be fully deterministic. We did not cap at 0 in order to allow for a small amount of accumulated floating point errors.

Initial Seeds. To obtain the initial seeds we took the set of 382 benchmarks in OpenQASM 3 form from the `mqtbench`[7] benchmarks. We minimised them down to a subset of 22 testcases that covered all edge cases found in the complete set using `py-afl-cmin`[8]. We modified these testcases to minimise the number of unique variable identifiers and added these to the grammar in order to increase the probability of generating programs free from undeclared variable usage.

Fuzzing Campaigns. Many short campaigns were run whilst creating and debugging the pipeline, often these produced large numbers of crashes due to unhandled exceptions during the parsing phase. We noted down each unique crash type and decided whether it was handled sufficiently gracefully or not; we added catches within our fuzzing harness for those exceptions that we believe were handled gracefully within the quantum library, as our aim was to test the functionality of the simulators rather than our tools ability to create valid QASM programs. In some cases we found exceptions during the parsing phase that were not well handled and could benefit from providing more context to the user. We report results from our longest run campaign which was performed on a 2023 Macbook Air with M2 processor and 8 GB RAM; this ran for 106 h on a single core.

3   Results

Our longest fuzzing campaign resulted in 407k executions (and thus approximately as many unique QASM programs), generating a corpus of 2,327 testcases (from an original 22), 139 saved crashes and 153 saved hangs; note that these are AFL++ statistics where only inputs that cover new functionality are saved, thus the resultant testcases significantly increase the diversity of the orignal 22 tests.

We discovered that 4 types of bugs were responsible for all 139 crashes; they are listed in order of severity in the rest of this paragraph. One bug filed to Qiskit maintainers ended up being an error in the specification for the OpenQASM 3

---

[7] https://www.cda.cit.tum.de/mqtbench/.

[8] https://github.com/jwilk/python-afl/blob/master/py-afl-cmin.

standard library (`stdgates.inc`); this directly led to us publicly filing a report in Braket and privately alerting Quantastica to a potential error in their simulator. Qiskit's standard library implementation has now been fixed, and Braket's simulator too. Additionally, we reported a significant performance issue in Braket, which has now been fixed; and a crash in the Quantastica simulator which has also been fixed. We describe these bugs further:

Braket. In Braket we found that the QASM interpreter made certain assumptions about the available attributes of some objects; this resulted in uncaught exceptions that provided no context about where the error occurred. In most cases, these were invalid programs and our only concern was that the error message was less helpful than others which provided context about which line and token the interpreter failed on. It is certainly possible to rationally argue that this is a whole family of bugs (though we do not), and we have witnesses to eight different crash locations.

<Bug BRAKET1>: In one of these cases the interpreter did not conform to the OpenQASM 3 specification, and after reporting we are assured that this will be addressed in a future release.

<Bug BRAKET2>: We filed another bug report whereby simulating relatively simple quantum circuits of 14 qubits resulted in the Python interpreter being killed due to running out of memory. This was due to an error in the implementation of the `gphase` builtin instruction, and has now been fixed by a maintainer; after which the simulator could comfortably handle the same circuits with 25 or more qubits.

Qiskit. For consistency between platforms, we chose to manually expand the `include "stdgates.inc"` statement using a simple string replacement with the file definition from the original OpenQASM 3 specification publication [2]. All three simulators provide inbuilt definitions for the standard gates, however, as a direct result of forcing them to have to generate the definitions directly from QASM we discovered that Qiskit's output probability distributions diverged from the other simulators for testcases involving use of the `sx` gate.

<Bug QISKIT1>: After triaging and discovering that the divergence only occurred when manually specifying the `sx` gate definition in QASM (rather than relying on the built-in definition), we decided to file an issue. The maintainers narrowed it down to one line: `gate sx a { pow(1/2) @ x a; }`. According to the OpenQASM 3 specification, dividing two integer literals should use integer division resulting in 1/2 resolving to 0; whereas Braket performed float division resulting in a value of 0.5. Ultimately it was decided that the error was in the original `sx` gate definition as provided in the specification and it should instead have either written the fraction as 1.0/2 (or equivalent) or used 0.5. While this meant that the `stdgates.inc` (the equivalent of OpenQASM 3's standard library) would need correcting, it also meant that Braket has an implementation error in applying floating point division where integer division should be used. Quantastica's behaviour aligned with Qiskit due to the OpenQASM 2 code being generated by Qiskit re-exporting the circuit that it had produced – after realising the divergent behaviour in Braket, we manually constructed a simple testcase to

check Quantastica and found that it too uses floating point division. We filed an issue with Braket to alert them to the issue, and privately informed the Quantastica maintainers of the finding, though as the OpenQASM 2 specification is less rigorous and does not directly specify integer division, we are not treating this as a bug. The Qiskit maintainers have merged a fix to the stdgates.inc standard library file, and Braket maintainers have fixed the floating point division error.

Quantastica. <Bug QUANT1>: The Quantastica simulator does some undocumented processing of register identifiers, causing some of our generated testcases to throw an out-of-heap-memory error. Identifiers that trigger this bug are also generated by Qiskit when exporting to OpenQASM 2 – this is not a highly improbable bug to encounter. The online tools quantum-circuit and q-convert run this JavaScript code in the browser, and attempting to parse code containing the bad identifiers results in a hang rather than a crash. As this bug could be used in a denial-of-service attack, we first emailed the library maintainer to ensure that no server-side applications could be targeted and only filed a publicly visible issue once we had assurance that all applications were run client-side.

3.1 Effectiveness of the Differential Testing Approach

Our search-based fuzz testing approach revealed multiple crashes and hangs. We can directly attribute the differential testing approach to the discovery of <Bug BRAKET2> causing crashes for gphase instructions with 14 qubits, and <Bug QISKIT1> where floating point division was incorrectly used in place of integer division. As many of the auto-generated circuits had large numbers of qubits, many testcases crashed or were killed, so if we were just testing Braket alone we may not have realised there was an issue. It was only after we spotted that Qiskit correctly handled one particular testcase that Braket crashed on that we decided to investigate further and discovered that there was probably an implementation issue. In the case of the division bug, it was as a direct result of the Jensen-Shannon divergence bounding assert being failed that this was detected. Testing any of the simulators on their own could have only discovered this with an appropriate oracle – which we do not have.

4 Conclusions

We proposed to use search-based differential testing to check validity of quantum program simulators. In particular, we used grammar-aware fuzzing to generate valid programs, which were then fed into different quantum simulators. Our results from over 400k executions show that our approach is useful in finding real bugs in such software.

Funding. We thank the ERC Advanced Grant no. 741278 and UK EPSRC Grant no. EP/S022503/1.

References

1. de la Barrera, A.G., de Guzmán, I.G.R., Polo, M., Piattini, M.: Quantum software testing: state of the art. J. Softw. Evol. Process. **35**(4), e2419 (2023)
2. Cross, A., et al.: OpenQASM 3: a broader and deeper quantum assembly language. TQC **3**(3), 1–50 (2022)
3. Fioraldi, A., Maier, D., Eißfeldt, H., Heuse, M.: AFL++: combining incremental steps of fuzzing research. In: WOOT Workshop. USENIX Association (2020)
4. Menéndez, M., Pardo, J., Pardo, L., Pardo, M.: The Jensen-Shannon divergence. J. Franklin Inst. **334**(2), 307–318 (1997)
5. Wang, J., Zhang, Q., Xu, G.H., Kim, M.: QDiff: differential testing of quantum software stacks. In: ASE Conference, pp. 692–704. IEEE (2021)

GreenStableYolo: Optimizing Inference Time and Image Quality of Text-to-Image Generation

Jingzhi Gong[1(✉)] , Sisi Li[2(✉)] , Giordano d'Aloisio[3(✉)] ,
Zishuo Ding[4(✉)] , Yulong Ye[5(✉)] , William B. Langdon[6(✉)] ,
and Federica Sarro[6(✉)]

[1] Loughborough University, Loughborough, UK
j.gong@lboro.ac.uk
[2] Beijing University of Posts and Telecommunications, Beijing, China
sisili@bupt.edu.cn
[3] Universitá degli Studi dell'Aquila, L'Aquila, Italy
giordano.daloisio@graduate.univaq.it
[4] University of Waterloo, Waterloo, Canada
zishuo.ding@uwaterloo.ca
[5] University of Birmingham, Birmingham, UK
yxy382@student.bham.ac.uk
[6] University College London, London, UK
{w.langdon,f.sarro}@ucl.ac.uk

Abstract. Tuning the parameters and prompts for improving AI-based text-to-image generation has remained a substantial yet unaddressed challenge. Hence we introduce GreenStableYolo, which improves the parameters and prompts for Stable Diffusion to both reduce GPU inference time and increase image generation quality using NSGA-II and Yolo. Our experiments show that despite a relatively slight trade-off (18%) in image quality compared to StableYolo (which only considers image quality), GreenStableYolo achieves a substantial reduction in inference time (266% less) and a 526% higher hypervolume, thereby advancing the state-of-the-art for text-to-image generation.

Keywords: SBSE · ANN · GenAI · Text2Image · Stable Diffusion · Yolo

1 Introduction

In recent years Generative Artificial Intelligence (GenAI) has emerged as a powerful approach encompassing various techniques that enable machines to generate new content, such as text [13], images [19], and videos [14]. Particularly, image generation and text-to-image synthesis have garnered significant attention due to their potential in bridging the gap between textual descriptions and visual representations [18]. It enables systems to understand and interpret human language and automatically translate it to a visually meaningful way, facilitating

G. Jahangirova and F. Khomh (Eds.): SSBSE 2024, LNCS 14767, pp. 70–76, 2024.
https://doi.org/10.1007/978-3-031-64573-0_7

tasks such as generating accompanying images for books [4], generating product images for advertising [22], and inspiring artists to create new forms of art [12].

However, achieving optimal performance in image generation tasks involves fine-tuning various aspects of a GenAI model, such as the number of inference steps, positive and negative prompts [1,3]. These parameters play a crucial role in determining the quality and efficiency of the generated images and tuning these parameters is essential to unlock the full potential of image generation models [3,15]. At the same time, GenAI model are energy demanding and largely contribute to the increased CO_2 emissions [8,20].

Berger et al. [3] proposed a search-based approach, dubbed StableYolo, to optimize the image quality of Stable Diffusion by assessing image quality using Yolo [17]. However, they their approach do not take into account the aspect of inference time, which is a cornerstone for both ensuring user experience and minimizing the energy consumption of GenAI models. Especially in real-world scenarios, where responsiveness and energy efficiency are vital, addressing this aspect is vital for the widespread adoption of such models [5,11,20,21].

To address this gap, we present GreenStableYolo, a novel approach that addresses the challenge of optimizing the trade-off between inference time and image quality using a search-based multi-objective optimization method, namely Non-dominated Sorting Genetic Algorithm (NSGA-II) [6]. We provide initial empirical evidence that by using GreenStableYolo Stable Diffusion achieves a satisfactory equilibrium between inference time and image quality, making it suitable for real-world applications where both factors play a crucial role.

In a nutshell, the key contributions of this work include:

- The development of a novel system that seeks for an optimal trade-off between inference time and image quality by optimizing the prompts and parameters for Stable Diffusion, dubbed GreenStableYolo;
- Empirical evidence on the effectiveness of GreenStableYolo in achieving significantly less inference time and higher hypervolume compared to StableYolo, thereby advancing the state-of-the-art multi-objective optimization for text-to-image generation;
- A comprehensive analysis to understand the influence of different parameters and prompts on both inference time and image quality in Stable Diffusion.

2   Related Work

To improve image generation quality, Berger et al. [3] were the first to propose the use of a Genetic Algorithm (GA) able to simultaneously tune the prompt and parameters of Stable Diffusion. Magliani et al. [15] use GA to find the best diffusion parameters for automated image retrieval from a dataset. While some research [5,11] has been carried out to optimize inference time, from hardware design to model architecture, there has been limited work focusing on optimizing the prompts and parameters. Our work builds upon previous work by considering both inference time and image quality as optimization objectives.

3   Methodology

To mitigate the aforementioned challenge, we propose GreenStableYolo, a novel multi-objective search-based approach that, given a text prompt for image generation, searches for the optimal parameters that can strike a balance between: (1) *Inference time*, which is measured by the GPU time taken for the execution of the StableDiffusionPipeline; and (2) *Image quality*, which is determined by performing object recognition with Yolo, then selecting objects that match the input prompt, and computing their average probabilities [3].

NSGA-II Optimization Algorithm. To simultaneously enhance image quality and reduce inference time, we leverage NSGA-II, a well-known and efficient multi-objective evolutionary algorithm [10,16]. Specifically, NSGA-II works as follows: (1) Initialize a population with N individuals; (2) Perform crossover and mutation operations, generating an offspring population denoted as P_o; (3) Reassemble the parent population P_{t-1} and P_o into a temporary population with the size of $2N$, and formulate individuals into i non-inferior frontier through fast non-dominating sorting; (4) Select N individuals from the temporary population to form the next population for the tth iteration, denoted as P_t. (5) Repeat steps (2)–(4) until the termination condition is met; and (6) The algorithm ends up and returns the current Pareto-Optimal set.

Selected Parameters. To make a straightforward comparison with StableYolo, we adopt the same settings as used by Berger et al. [3]. Specifically, the following parameters and prompts are tuned and searched with NSGA-II: (1) *Inference steps* (1 to 100): the AI's image generation iterations; (2) *Guidance scale* (1 to 20): the impact of the prompt on image generation; (3) *Guidance rescale* (0 to 1): rescales the guidance factor to prevent over-fitting; (4) *Seed* (1 to 512): randomization seed; (5) *Positive prompt*: used to describe images and improve their details, e.g., "photograph", "color", and "ultra real" [2]; and (6) *Negative prompt*: avoided description during image generation, e.g., "sketch", "cropped", and "low quality" [?]

4   Evaluation

To evaluate our proposal, we address the following research questions (RQs):
➤ **RQ1:** To what extent can GreenStableYolo improve image quality and inference time compared with StableYolo?
➤ **RQ2:** How do parameters/prompts of Stable Diffusion influence the inference time for image generation?
➤ **RQ3:** How do parameters/prompts of Stable Diffusion influence the quality of the generated images?

Experimental Setup. To ensure a fair evaluation of the optimization effectiveness, we employed the same hyperparameter setup as StableYolo for NSGA-II, e.g., the population size was set to 25, the number of generations was set to 50, and both the mutation rate and crossover rate were set to 0.2. We selected the weights of 0.001 for image quality and -1000 for inference time based on empirical investigation of different weight combinations. In addition, we used Stable Diffusion version v2 and Yolo version v8. To assess variability, we evaluated each model 15 times using different random seeds, focusing solely on the prompt "two people and a bus" due to time constraints. Any future studies can explore additional prompts. All experiments were conducted on a virtual machine hosted on Google Colaboratory, with an NVIDIA Tesla T4 GPU with 16 GB of RAM.

RQ1 Results. Figure 1 presents the performance comparisons between GreenStableYolo and StableYolo. Specifically, Fig. 1a reveals that GreenStableYolo achieves an average inference time of 9.4 s with an interquartile range (IQR) of 4.7 s. Conversely, StableYolo exhibits an average inference time of 25.0 s, which is 1.66 times higher than GreenStableYolo, with an IQR of 9.1 s. That is, GreenStableYolo generates images much faster.

This improvement in inference time comes at a slight cost to image quality. As illustrated in Fig. 1b, GreenStableYolo experiences approximately an average degradation of 0.18 points in image quality. We also compute the hypervolume [9] for both models for a more comprehensive comparison[1]. Figure 1c presents the hypervolume values with the reference point set as (1, 50000), where GreenStableYolo achieves an average hypervolume of 29074.11, surpassing StableYolo's score of 4642.17 by 5.26 times. *This substantial difference demonstrates the clear dominance of GreenStableYolo over StableYolo in this two-objective optimization problem for text-to-image generation.*

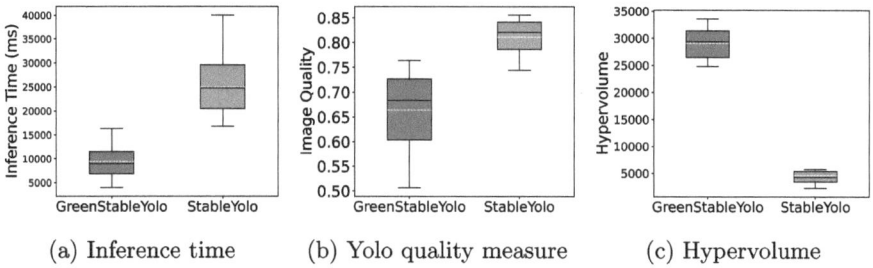

| (a) Inference time | (b) Yolo quality measure | (c) Hypervolume |

Fig. 1. Comparison of GreenStableYolo and StableYolo on 15 independent runs

RQs2–3 Results. To investigate RQs2–3, we followed previous work [7] and built two Random Forest regression models using `scikit-learn`. The features of these models include the number of iteration steps, guidance scale, guidance rescale, positive prompts, and negative prompts (excluding the random seed). The target variables are inference time and image quality score, respectively. We

---

[1] Hypervolume is a fundamental metric used in multi-objective optimization problems that indicates the dominance of a solution in the objective space.

use the `RandomizedSearchCV` function from `scikit-learn` to find the optimal hyperparameters during model training. The `feature_importances_` function is then used to compute the importance of each parameter and prompt based on the Mean Decrease Impurity (MDI), a.k.a. as Gini importance. To ensure reliability, we repeat this process 10 times.

Figures 2a and b present the calculated importance of parameters and prompts based on the mean decrease in impurity, with respect to inference time and image quality scores, respectively. As shown in Fig. 2a, the number of *inference steps* emerges as a significant factor affecting inference time. This is expected, as more steps involve more computations, thereby resulting in higher inference time. Meanwhile, for image quality (Fig. 2b), parameters like *guidance rescale* and *positive prompts* play a relatively more critical role.

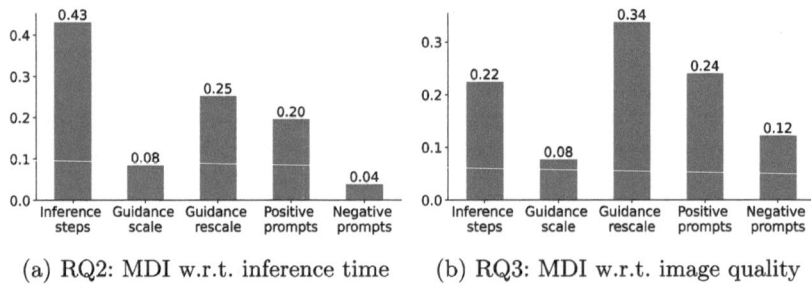

(a) RQ2: MDI w.r.t. inference time   (b) RQ3: MDI w.r.t. image quality

Fig. 2. Parameters/prompts importance based on the mean decrease in impurity

These findings confirm the value of our work: *increased computational resources do not necessarily translate to better image quality; instead, appropriate model parameter settings are more crucial.* This highlights the importance of identifying optimal parameter combinations during model inference to balance computational efficiency and output quality.

Threats to Validity. The limited exploration of prompts, the randomness in the optimization process, and the specific configuration for NSGA-II may introduce internal threats. Besides, external threats may include the choice of the GenAI model, the noise when measuring the inference time, and the evaluation of image quality based on object recognition using Yolo.

5   Conclusion

In GenAI text-to-image, achieving images of high-quality is often not the only important aspect to consider, as inference time, which directly impacts user experience and energy consumption, also plays a critical role. In this work we introduced GreenStableYolo, the first approach leveraging NSGA-II to strike an optimal trade-off between these two objectives for Stable Diffusion. Experimental comparisons with StableYolo demonstrate that GreenStableYolo achieves significantly reduced inference time while maintaining a relatively high image quality. Future research can expand upon our evaluation by incorporating alternative initial prompts, optimizing different performance metrics such as energy consumption, and broadening to other GenAI systems such as DALL-E, ImageFX, or Midjourney.

Availability. Repository available at https://github.com/gjz78910/GreenStable Yolo.

Acknowledgments. Work supported by the ERC grant no. 741278.

References

1. Stable diffusion documentation. https://huggingface.co/docs/diffusers/api/pipelines/stable_diffusion/text2img
2. Stable diffusion photorealistic prompts. https://prompthero.com/stable-diffusion-photorealistic-prompts
3. Berger, H., et al.: StableYolo: optimizing image generation for large language models. In: Arcaini, P., Yue, T., Fredericks, E.M. (eds.) Search-Based Software Engineering, SSBSE 2023. LNCS, vol. 14415, pp. 133–139. Springer, Cham (2024). https://doi.org/10.1007/978-3-031-48796-5_10
4. Bruens, J.D., Meissner, M.: Do you create your content yourself? Using generative artificial intelligence for social media content creation diminishes perceived brand authenticity. J. Retail. Consum. Serv. **79**, 103790 (2024)
5. Cao, Y., et al.: A comprehensive survey of AI-generated content (AIGC): a history of generative AI from GAN to ChatGPT. arXiv arXiv:2303.04226 (2023)
6. Deb, K., Agrawal, S., Pratap, A., Meyarivan, T.: A fast and elitist multiobjective genetic algorithm: NSGA-II. IEEE Trans. Evol. Comput. **6**(2), 182–197 (2002)
7. Ding, Z., Chen, J., Shang, W.: Towards the use of the readily available tests from the release pipeline as performance tests: are we there yet? In: ICSE 2020, pp. 1435–1446 (2020)
8. Georgiou, S., Kechagia, M., Sharma, T., Sarro, F., Zou, Y.: Green AI: do deep learning frameworks have different costs? In: Proceedings of the ICSE 2022, pp. 1082–1094 (2022)
9. Guerreiro, A.P., Fonseca, C.M., Paquete, L.: The hypervolume indicator: computational problems and algorithms. ACM Comput. Surv. **54**(6), 1–42 (2021)
10. Ji, B., Huang, H., Yu, S.S.: An enhanced NSGA-II for solving berth allocation and quay crane assignment problem with stochastic arrival times. IEEE Trans. Intell. Transp. Syst. **24**, 459–473 (2023)
11. Kim, S., et al.: Full stack optimization of transformer inference: a survey. arXiv arXiv:2302.14017 (2023)
12. Ko, H., et al.: Large-scale text-to-image generation models for visual artists' creative works. In: Intelligent User Interfaces, IUI 2023, pp. 919–933 (2023)
13. Li, J., Tang, T., Zhao, W.X., Wen, J.R.: Pretrained language models for text generation: a survey. arXiv arXiv:2105.10311, 25 May 2021
14. Long, F., Qiu, Z., Yao, T., Mei, T.: VideoDrafter: content-consistent multi-scene video generation with LLM. arXiv arXiv:2401.01256, 2 January 2024

15. Magliani, F., Sani, L., Cagnoni, S., Prati, A.: Genetic algorithms for the optimization of diffusion parameters in content-based image retrieval. In: ICDSC 2019 (2019)
16. Rashid, T.A., et al.: NSGA-II-DL: metaheuristic optimal feature selection with deep learning framework for HER2 classification in breast cancer. IEEE Access **12**, 38885–38898 (2024)
17. Redmon, J., Divvala, S.K., Girshick, R.B., Farhadi, A.: You only look once: unified, real-time object detection. In: CVPR 2016, pp. 779–788 (2016)
18. Reed, S.E., Akata, Z., Yan, X., Logeswaran, L., Schiele, B., Lee, H.: Generative adversarial text to image synthesis. In: ICML 2016, pp. 1060–1069 (2016)
19. Rombach, R., Blattmann, A., Lorenz, D., Esser, P., Ommer, B.: High-resolution image synthesis with latent diffusion models. In: CVPR, pp. 10674–10685 (2022)
20. Sarro, F.: Search-based software engineering in the era of modern software systems. In: IEEE International Requirements Engineering Conference, pp. 3–5 (2023)
21. Sarro, F.: Automated optimisation of modern software system properties. In: International Conference on Performance Engineering, ICPE 2023, pp. 3–4 (2023)
22. Yu, T., Yang, X., Jiang, Y., Zhang, H., Zhao, W., Li, P.: TIRA in Baidu image advertising. In: International Conference on Data Engineering, ICDE 2021 (2021)

Danger is My Middle Lane: Simulations from Real-World Dangerous Roads

Antony Bartlett[(✉)] and Annibale Panichella

Delft University of Technology, Delft, The Netherlands
{a.j.bartlett,a.panichella}@tudelft.nl

Abstract. Self-driving cars face significant testing challenges, constrained by high costs and limited test environments. This necessitates innovative approaches to simulation-based testing to improve deployment and evaluation efficiency. Our study introduces a novel methodology that leverages dangerous real-world road maps sourced from Google Earth as the initial seed for generating driving scenarios. In contrast, traditional approaches use randomly generated seeds/maps to initialize the search process. We systematically adjust road points by evolving these maps to induce out-of-bounds (OOB) errors. Our preliminary results demonstrate a significant improvement in generating failing scenarios, when using real-world maps as seeds compared to random seeds/maps. Specifically, the evolved real-world maps are more likely to be valid (e.g., not self-intersecting) and have a higher incidence of OOB failures. This work opens avenues for further research into optimizing scenario generation for broader applications in autonomous systems testing.

Keywords: Cyber-Physical Systems · Simulation-based Testing · Evolutionary Search

1 Introduction

Driving remains perilously unsafe, evidenced by over one million road fatalities annually [8], with road injury leading as the cause of death among individuals aged 5 to 29 in 2019 according to the World Health Organization. Autonomous vehicles emerge as a potential solution to enhance road safety, with companies like Waymo, Cruise, and Tesla building and running these vehicles across the USA. However, autonomous driving implementation faces challenges, underscored by instances of malfunctions and accidents [6,9]. Therefore, considerable testing is required to ensure autonomous vehicles function safely and correctly.

Simulation-based testing offers a promising alternative, enabling the efficient generation and execution of diverse driving scenarios without the need for the hardware-in-the-loop. Existing frameworks like Carla[1] and BeamNG[2] offer

---

[1] https://github.com/carla-simulator/carla.
[2] https://www.beamng.tech/.

© The Author(s), under exclusive license to Springer Nature Switzerland AG 2024
G. Jahangirova and F. Khomh (Eds.): SSBSE 2024, LNCS 14767, pp. 77–83, 2024.
https://doi.org/10.1007/978-3-031-64573-0_8

sophisticated environments that simulate real-world dynamics, weather, and traffic patterns essential for rigorous testing. Recent advancements in autonomous driving testing have explored various strategies for scenario generation aimed at challenging specific features, such as lane-keeping. The SBFT 2023 competition [3] showcased innovative methods, such as combining evolutionary search with reinforcement learning [7], leveraging Wasserstein generative adversarial networks [11], and employing Extended Finite State Machines (EFSMs) [4].

However, a prevalent methodology among driving scenario generators has been the use of randomly generated initial road maps (random seeds) [1,2], optimized iteratively to induce safety requirement violations, such as driving off lane. While this approach has its merits, it often produces many invalid scenarios [3] or requires various iterations to achieve a single safety violation.

In this paper, we embark on a preliminary investigation into using real-world maps as seeds for driving scenario generators. To achieve this, we sampled 25 1-km long segments of roads, identified as some of the most dangerous globally[3]. These roads are subsequently evolved through multi-objective evolutionary algorithms [1,2]. Although these maps do not lead directly to failing scenarios, we conjecture they provide better starting points (seeds) for evolution.

We developed two variants of multi-objective road generators, leveraging the framework provided by the latest SBFT tool competition, which integrates BeamNG as a simulation environment. These variants diverge in their initial seeding strategy; one employs real-world maps, while the other utilizes randomly generated roads. Targeting BeamNG's built-in AI to evaluate the lane-keeping feature, we aimed to assess the efficacy of each seeding approach in generating challenging scenarios, OOB errors in this example. Our preliminary results demonstrate that the real-world map variant produced a significantly higher number of failing scenarios and generated more valid roads (e.g., non-self-intersecting), as evidenced by large effect sizes in our statistical analysis. This result opens new avenues for research, particularly in enhancing the realism and applicability of simulated testing environments for autonomous systems.

2   Seeding from Real-World Maps

To validate our conjecture, we have implemented a multi-objective (1+1) Evolutionary Algorithm (EA) with an archive[4], utilizing the framework provided by the SBFT 2024 CPS competition tool[5]. We use an algorithm that evolves only one solution/map at a time, as recommended in related literature for problems with expensive-to-evaluate solutions [1]. (1+1) EA starts with an initial road map, either randomly generated (first variant) or seeded from a real-world map pool (second variant). After execution, the initial map is stored in the archive, which keeps track of the non-dominated solutions across the generations. At each generation, a new solution is obtained by either (1) mutating a solution from

---

[3] https://www.dangerousroads.org.

[4] https://doi.org/10.5281/zenodo.11221209.

[5] https://github.com/sbft-cps-tool-competition/cps-tool-competition.

the archive or (2) by sampling a completely new road (randomly generated or sampled from the real-world map pool). The new/mutated map is executed and compared with the existing solutions in the archive. We detail the main elements of the search algorithm in the following paragraphs.

Real-World Pool. Our approach utilizes real-world roads from Google Earth[6], in the form of Keyhole Markup Language[7] (KML) files, an XML standard notation for geographic annotations and visualization for Earth browsers. Specifically, we sampled 25 real-world maps from an online database of dangerous roads from around the world[8]. For this, we developed a pre-processing script that extracts a kml file from Google Maps given the road information from the database.

Solution Encoding. We use the same encoding as the SBFT CPS competition tool: a road map is characterized by a list of control nodes/points $P = \{(x_1, x_2), \ldots (x_n, y_n)\}$, where x_i and x_j denote the x and y coordinates of the i-th node in the map. The road is generated by interpolating the control points using the cubic spline interpolation.

Initialization and Sampling. In the initial phase and through the search, new road maps are sampled. We implemented two sampling variants: (1) real-world map sampling and (2) random sampling. In the former variant, the real-world maps are a randomly selected *seeding pool* described above. In the latter variant, a new road is generated by randomly generating control nodes and their positions. The number of nodes is randomly selected between 7 and 10, and the position of each node is randomly selected within the 1km road length.

Search Objectives. Our algorithm optimizes three objectives: (1) the total out-of-bound (OOB) percentage, (2) the maximum speed of the vehicle, and (3) the maximum steering angle of the vehicle. The OOB percentage measures the percentage of the vehicle that is out-of-bounds during the simulation. A zero percentage indicates that the vehicle is exactly within one of the lane boundaries (lines); a negative value measures the distance to the lane boundaries; finally, a positive value indicates that the vehicle is out-of-bounds and the percentage value indicates the amount/percentage of the vehicle that is out-of-bounds.

Archive and Non-dominated Selection. Non-dominated solutions are stored in an archive across test generations. Therefore, each newly generated/mutated individual is compared with the individuals stored in the archive. If the new individual is non-dominated, such that the three objectives have moved closer to the maximum, it is added to the archive. If the new individual dominates an individual in the archive, the latter is removed from the archive. Otherwise, if an individual in the archive dominates the new solution, the latter is discarded.

Reproduction. The search algorithm randomly selects one of the non-dominated solutions from the archive. The new road map is generated by applying three types of mutations: (1) adding a new control node, (2) removing a

---

[6] https://earth.google.com.

[7] https://en.wikipedia.org/wiki/Keyhole_Markup_Language.

[8] https://www.dangerousroads.org.

control node, or (3) changing the position of a control node. Each mutation type is performed with a probability of $1/3$.

Test Execution. All the generated roads are designed to work within a 1km map grid. Before a test is simulated, the tool checks that the generated road fits correctly within the map boundaries. This is to eliminate any possible false negatives, such as when the car starts immediately outside the map boundary and is categorized as out-of-bounds (OOB). The road is then validated within the SBFT'24 CPS framework for its viability. Numerous checks are carried out, such as ensuring the road does not intersect or have too sharp corners. If the road is now considered viable, then the simulation is executed. During the simulation, the road and a car will be loaded within BeamNG, and the states will be monitored to check that our car remains within the scope of our OOB parameters.

Final Remark. Seeding strategies have been investigated in evolutionary testing for unit-level test case generation. In the context of autonomous driving, our work is the first to explore using real-world maps as seeds for evolving driving scenarios. The closest work to ours is the approach by Gambi et al. [5] and Nguyen et al. [10]. The former approach generates driving scenarios from police incident reports. They use natural language processing (NLP) methods to reconstruct driving scenarios from natural language sentences. Nguyen et al. [10] also create driving scenarios starting from existing maps stored in OpenDrive format. Compared to these related works, our approach (1) extracts the road maps from Google Earth and (2) evolves these maps since (based on our preliminary results) they do not directly lead to test failures without any evolution.

3   Preliminary Study

The goal of our preliminary study is to answer the following research question:

RQ1: *How effective is real-world road seeding compared to random seeding when generating driving scenarios?*

We compare two variants of (1+1) EA, which differ w.r.t. seeding: one employs real-world maps, while the other utilizes randomly generated roads as described in the previous section. For testing, we target BeamNG's built-in AI to evaluate the lane-keeping, along with features included in the SBFT CSP tool. For comparison we run each variant 8 times, comparing the results w.r.t (1) number of failed test scenarios due to OOB violations, (2) number of valid test roads as reported by the validity checkers of the tool competitions (e.g., non-intersecting). For statistical analyses, we use the statistical tests recommended by the literature [1,3]. We use the Wilcoxon rank sum test to assess the significance of the difference (if any) with a p-value threshold set to 0.05. Additionally, we also use the Vargha-Delaney ($\hat{A}_{12}$) as the measure for the effect size.

Parameter Settings. When running the tests, we allocated a time budget of 28800 s (8 h). A test fails if the driving vehicle is 50% out-of-bounds. We ensure that when the vehicle is no more than 50% out-of-bounds, the test is still considered as passing. Finally, we set a maximum speed of 85 km/h for our vehicle and the *aggressive* driving style set to 1.0, as suggested by the tool competition when using BeamNG. Such a value sets the autonomous agent with a balanced driving style, i.e., not too aggressive or not too cautious.

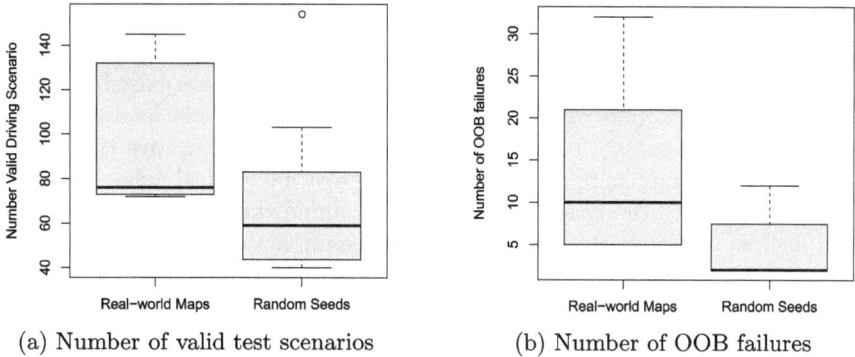

(a) Number of valid test scenarios   (b) Number of OOB failures

Fig. 1. Number of OOB failures

Preliminary Results. Figures 1b and a depict the boxplots of the number of OOB failures and valid test scenarios, respectively. The results show that the real-world seed maps perform considerably better than the random-seeding counterpart in terms of both the number of failing scenarios and the number of generated valid roads. On average, real-world seeding leads to five times more failing driving scenarios (the median values being 10 vs. 2) and 30% more valid roads compared to random seeding. When considering the number of OOB failures, the Wilcoxon rank sum test confirms the significance of the difference with a p-value < 0.01 and a large effect size ($\hat{A}_{12} = 0.80$). The statistical difference also holds for the number of valid scenarios, with a p-value $= 0.02$ and $\hat{A}_{12} = 0.76$ (large effect). Based on these findings, we can conclude that real-world seeding performs considerably better than its random counterpart. This supports our initial intuition that real-world maps provide better starting points for generating challenging driving scenarios.

Threats to Validity. The obvious threat to the validity of this work is the size of our study, as we target only one autonomous system (BeamNG). For internal validity, a larger number of test runs could have been considered. However, this would incur a significant computational cost as the current study required 128 h of test execution. As this is a preliminary study and given the large effect size observed in our results, we are confident that further executions would result in a similar outcome. The other threat is thus external, as this preliminary study only considered random seeding as the baseline. Replicating these results in a larger study with other road generators would aid in minimizing external threats.

4   Conclusions and Future Work

Our study reveals that real-world roads, with their inherent complexity due to longer lengths and more frequent turns, provide a more robust and more effective starting point for generating driving scenarios compared to randomly generated roads. Additionally, having more valid test case scenarios makes it more likely that evolution can find failing mutations. This preliminary work serves as a baseline for the inclusion of real-world elements in simulation-based testing. It is not simply the corners and length of these roads that would lead them to be dangerous in the real world, but other factors as well. To improve this work in the future, we would like to investigate the usage of the vertical axis, adding hills and blind verges to the roads. We will also examine the effect of considering a larger sample of real-world roads, as well as the impact of varying the length of the road maps. More aspects of the mutation will also allow us to experiment with other many-objective search techniques. This will extend the experiment by comparing the efficacy of these various algorithms for seed mutation.

Acknowledgements. This work has been supported by the European Union's Horizon 2020 Research and Innovation Programme under grant agreement No. 957254, project COSMOS (2021).

References

1. Abdessalem, R.B., Panichella, A., Nejati, S., Briand, L.C., Stifter, T.: Testing autonomous cars for feature interaction failures using many-objective search. In: Proceedings of the 33rd ACM/IEEE International Conference on Automated Software Engineering, pp. 143–154 (2018)
2. Ayerdi, J., Arrieta, A., Illarramendi, M.: Roadsign at the SBFT 2023 tool competition cyber-physical systems track. In: 2023 IEEE/ACM International Workshop on Search-Based and Fuzz Testing (SBFT), pp. 37–38. IEEE (2023)
3. Biagiola, M., Klikovits, S., Peltomäki, J., Riccio, V.: SBFT tool competition 2023 - cyber-physical systems track. In: 2023 IEEE/ACM International Workshop on Search-Based and Fuzz Testing (SBFT), pp. 45–48 (2023)
4. Ferdous, R., Hung, C.K., Kifetew, F., Prandi, D., Susi, A.: EvoMBT at the SBFT 2023 tool competition. In: 2023 IEEE/ACM International Workshop on Search-Based and Fuzz Testing (SBFT), pp. 59–60 (2023)
5. Gambi, A., Huynh, T., Fraser, G.: Generating effective test cases for self-driving cars from police reports. In: Proceedings of the 2019 27th ACM Joint Meeting on European Software Engineering Conference and Symposium on the Foundations of Software Engineering, pp. 257–267 (2019)
6. Guardian, T.: Cruise recalls all self-driving cars after Grisly accident and California ban. The Guardian. https://www.theguardian.com/technology/2023/nov/08/cruise-recall-self-driving-cars-gm
7. Humeniuk, D., Khomh, F., Antoniol, G.: RIGAA at the SBFT 2023 tool competition - cyber-physical systems track. In: 2023 IEEE/ACM International Workshop on Search-Based and Fuzz Testing (SBFT), pp. 49–50 (2023)
8. United Nations: With 1.3 million annual road deaths, un wants to halve number by 2030. United Nations

9. Neal, A.L.: I tried tesla's full self-driving mode. It nearly crashed my car. TC Palm. https://eu.tcpalm.com/story/opinion/columnists/2024/04/01/tesla-full-self-driving-test-near-rear-end-collision-florida-vero-beach/73170733007/

10. Nguyen, V., Huber, S., Gambi, A.: SALVO: automated generation of diversified tests for self-driving cars from existing maps. In: 2021 IEEE International Conference on Artificial Intelligence Testing (AITest), pp. 128–135. IEEE (2021)

11. Winsten, J., Porres, I.: WOGAN at the SBFT 2023 tool competition - cyber-physical systems track. In: 2023 IEEE/ACM International Workshop on Search-Based and Fuzz Testing (SBFT), pp. 43–44 (2023)

Author Index

G. Jahangirova and F. Khomh (Eds.): SSBSE 2024, LNCS 14767, p. 85, 2024.
https://doi.org/10.1007/978-3-031-64573-0